Sarjoo Prasad Yadav

Approximation Processes Involving Jacobi Series And Wavelets

AF138180

Sarjoo Prasad Yadav

Approximation Processes Involving Jacobi Series And Wavelets

Signals and Wavelets

LAP LAMBERT Academic Publishing

Impressum / Imprint

Bibliografische Information der Deutschen Nationalbibliothek: Die Deutsche Nationalbibliothek verzeichnet diese Publikation in der Deutschen Nationalbibliografie; detaillierte bibliografische Daten sind im Internet über http://dnb.d-nb.de abrufbar.
Alle in diesem Buch genannten Marken und Produktnamen unterliegen warenzeichen-, marken- oder patentrechtlichem Schutz bzw. sind Warenzeichen oder eingetragene Warenzeichen der jeweiligen Inhaber. Die Wiedergabe von Marken, Produktnamen, Gebrauchsnamen, Handelsnamen, Warenbezeichnungen u.s.w. in diesem Werk berechtigt auch ohne besondere Kennzeichnung nicht zu der Annahme, dass solche Namen im Sinne der Warenzeichen- und Markenschutzgesetzgebung als frei zu betrachten wären und daher von jedermann benutzt werden dürften.

Bibliographic information published by the Deutsche Nationalbibliothek: The Deutsche Nationalbibliothek lists this publication in the Deutsche Nationalbibliografie; detailed bibliographic data are available in the Internet at http://dnb.d-nb.de.
Any brand names and product names mentioned in this book are subject to trademark, brand or patent protection and are trademarks or registered trademarks of their respective holders. The use of brand names, product names, common names, trade names, product descriptions etc. even without a particular marking in this works is in no way to be construed to mean that such names may be regarded as unrestricted in respect of trademark and brand protection legislation and could thus be used by anyone.

Coverbild / Cover image: www.ingimage.com

Verlag / Publisher:
LAP LAMBERT Academic Publishing
ist ein Imprint der / is a trademark of
OmniScriptum GmbH & Co. KG
Heinrich-Böcking-Str. 6-8, 66121 Saarbrücken, Deutschland / Germany
Email: info@lap-publishing.com

Herstellung: siehe letzte Seite /
Printed at: see last page
ISBN: 978-3-659-48160-4

Approximation Processes Involving Jacobi Series and Wavelets

By

Sarjoo Prasad Yadav

"Every measurement is merely an approximation. Life and the universe depend upon approximation. And so, too, does technology." in 'Wavelet Analysis' by Resnikoff, H. L. and Wells, Jr. R. O.

Here,

A Lebesgue integrability controlled Signal is represented at first by Jacobi polynomials then by wavelets.

2

To

Grand Pa Late Sri Ram Manohar Yadav

CONTENTS

PREFACE

Mathematics, a word lying as a gem in every human heart, is the most luring subject on this planet. It is my beloved without which I might have not existed. It has mesmerized and delighted me all over my life. It is fate finding chance that I got involved in a branch of Mathematics not applauded by many influential and stalwart mathematicians. But a link with wavelets as elaborated in the treatise 'Wavelet Analysis' by Resnikoff and Wells and their company 'Aware, Inc.' have enchanted me to put forth my findings in a Monograph which is currently in our hands.

This monograph is a monorail assembled into five compartments as its five chapters. Chapter I is its engine provides necessary knowledge of 'Jacobi polynomials' while chapter II gives clues about instruments used in the preparation so that it rolls on the monorail line which are inequalities related to Jacobi polynomials. Chapters III, IV and V are comfortable sitting rooms. To enjoy the journey of study a basic knowledge of 'Summating Processes' is necessary.

Every chapter is in the form of section numbered by digits a.b to mean bth section in ath chapter. Formula (a.b.c) means cth formula in bth section of chapter a. Theorems, Lemmas and Remarks are numbered as a.b.c without brackets and they indicate cth item of bth section in chapter a.

Chapter III relates results on Jacobi polynomials and Jacobi series often called F-J expansion that are useful in approximation of functions by polynomials. Distinct summating operators are applied to find out bounds of many linear combinations of Jacobi polynomials. That is the reason that a reader is expected to have a prior knowledge of Cesàro and Nörlund means and other linear transforms of Jacobi series by Toeplitz matrices. Thus results in Chapter IV help one to follow the expression of a Lebesgue class function in terms of Jacobi polynomials.

Generalized translate of a signal defined in the interval $[0, \pi]$ plays vital role to ascend the challenging mounds of 'Approximation by Jacobi polynomials'. A general reading of IX chapter of the treatise 'Wavelet Analysis' is sufficient to understand our woven intricacy of "Approximation of a signal by wavelets" in the last chapter V of this monograph. I myself find in a mesh that how our study has claimed that a signal of diverging nature can be represented by wavelets, a tool to control communication skills (Chapter V Theorem 5.3.1). If we are doing something correct then we have actually opened a gate of dark continent where measure researches to be done are waiting for the good of the people at large.

7

CHAPTER I

INTRODUCTION

1.1 Jacobi Polynomials.

By a Polynomial or an Algebraic Polynomial we always mean an algebraic expression

(1.1.1a) $$1 + c_1 x + c_2 x^2 + c_3 x^3 + \ldots + c_n x^n$$

called polynomial in $x \in [a, b]$, ($-\infty \le a < b \le +\infty$) where c_i ($i = 0, 1, 2, 3, \ldots$) are real or complex numbers. If $c_n \ne 0$ then this polynomial is called 'of degree' n. This type of polynomial is basic in algebra from historical point of view so it is called classical.

Definition 1.1.1 *A classical algebraic polynomial denoted by $P_n^{(\alpha,\beta)}(x)$, orthogonal on $[-1, +1]$, with weight function $w(x) = (1 - x)^\alpha (1 + x)^\beta$, ($\alpha, \beta > -1$) is called Jacobi polynomial.*

The name 'Jacobi' is in the honor of German Mathematician Carl G. J. Jacobi (1804 – 1851), Professor at Konigsberg (1826 - 1843) and at the University of Berlin (1843 – 1851). Our study is mainly related to Jacobi series formed by the Jacobi polynomials. We, thus introduce main features of Jacobi polynomials. Its thorough study may be found in the treatise 'Orthogonal Polynomials' by G. Szegö[1].

Definition 1.1.2 *Jacobi polynomials $P_n^{(\alpha,\beta)}(x)$ satisfy the Hypergeometric Differential Equation of Gauss:*

(1.1.1b) $$x(1-x)\frac{d^2y}{dx^2} + [\alpha + 1 - (\alpha + \beta + 2)x]\frac{dy}{dx} + n(n + \alpha + \beta + 1)y = 0$$

for $y = constant . P_n^{(\alpha,\beta)}(x)$, ($n = 0, 1, 2, 3 \ldots$) *when* $\alpha, \beta > -1$.

The Jacobi polynomials can be represented by a Hypergeometric function F given by

(1.1.2a) $$P_n^{(\alpha, \beta)}(x) = \binom{n+\alpha}{n} F\left(-n, n + \alpha + \beta + 1; \alpha + 1; \frac{1-x}{2}\right)$$

Also,

(1.1.2b) $$\frac{d}{dx}\left\{P_n^{(\alpha, \beta)}(x)\right\} = \frac{1}{2}(n + \alpha + \beta + 1) P_{n-1}^{(\alpha+1, \beta+1)}(x)$$

(see Szegö[1] page 63). Thus we have

(1.1.3) $$P_n^{(\alpha, \beta)}(1) = \binom{n+\alpha}{n} \ne 0$$

(1.1.4) $$P_n^{(\alpha, \beta)}(-1) = (-1)^n \binom{n+\beta}{n}$$

and

(1.1.5) $$P_n^{(\alpha, \beta)}(-x) = (-1)^n P_n^{(\beta, \alpha)}(x)$$

Definition 1.1.3 *Jacobi polynomials* $P_n^{(\alpha,\ \beta)}(x)$ *satisfy the Rodrigue's formula*

(1.1.6) $\quad (1-x)^\alpha (1+x)^\beta \, P_n^{(\alpha,\ \beta)}(x) \ = \ \dfrac{(-1)^n}{2^n \, (n)!} \ \left(\dfrac{d}{dx}\right)^n \left\{(1-x)^{n+\alpha}(1+x)^{n+\beta}\right\}$

for arbitrary $\alpha,\ \beta$.

Following calculations are used for $\alpha, \beta > -1$:

(1.1.7) $\quad \int_{-1}^{+1}(1-x)^\alpha (1+x)^\beta \left\{ P_n^{(\alpha,\ \beta)}(x)\right\}^2 \, dx$

$$= \ \frac{2^{\alpha+\beta+1}}{2n+\alpha+\beta+1} \ \frac{\Gamma(n+\alpha+1)\Gamma(n+\beta+1)}{\Gamma(n+1)\Gamma(n+\alpha+\beta+1)} \ \equiv \ h_n^{(\alpha,\ \beta)}(say),$$

(see Szegö[1] page 68). The set $\{\, p_n\,(x)\}$ where

(1.1.8) $\qquad\qquad p_n\,(x) \ = \ \left\{h_n^{(\alpha,\ \beta)}\right\}^{-1/2} \, P_n^{(\alpha,\beta)}(x)\quad, \quad (n=0,1,2,3,\ldots)$

forms an orthonormal set on [- 1, +1] with respect to weight $w(x) \equiv (1-x)^\alpha(1+x)^\beta$ in $X = X_p^{\alpha,\ \beta} \cap L_1(w)$ which is collection of p-power $(1 \le p \le \infty)$ Lebesgue integrable with weight w and belongs to $L_1(w)$ functions. Condition $L_1(w)$ is required to expand the function into Jacobi series (see Szegö[1] discussion above (9.11.1).

Definition 1.1.4 $\quad$ *The Jacobi polynomials can also be defined by the recurrence relations*;

(1.1.9) $\ 2n\left(n+\alpha+\beta\right)(2n+\alpha+\beta-2)\, P_n^{(\alpha,\ \beta)}(x)$

$$= \left(2n+\alpha+\beta-1\right)\left\{\left(2n+\alpha+\beta\right)(2n+\alpha+\beta-2)\,x + \alpha^2 \ - \ \beta^2\right\} P_{n-1}^{(\alpha,\ \beta)}(x)$$

$$-\ 2\left(n+\alpha-1\right)\left(n+\beta-1\right)\left(2n+\alpha+\beta\right) P_{n-2}^{(\alpha,\ \beta)}(x)\ , (n=2,3,4,\ \ldots)\ ;$$

$$P_0^{(\alpha,\ \beta)}(x) \ = \ 1, \quad P_1^{(\alpha,\ \beta)}(x) \ = \ \tfrac{1}{2}\!\left(\alpha+\beta+2\right)x + \tfrac{1}{2}\!\left(\alpha-\beta\right)$$

This leads the concept that $P_n^{(\alpha,\ \beta)}(x)$ is a combination of $1, x, x^2, x^3, \ldots, x^n$ and has the form

(1.1.10) $\qquad\qquad P_n^{(\alpha,\ \beta)}(x) \ = \ c_0 + c_1\,x + c_2\,x^2 + c_3\,x^3 + \ldots + c_n\,x^n,\ \ (c_n \ne 0)$

where coefficients c_i (i = 0, 1, 2, ...) are real or complex numbers which depend upon n, α , β for n = 0, 1, 2, ... and α, $\beta > -1$. Jacobi polynomials $P_n^{(\alpha,\ \beta)}(x)$ are called Ultraspherical polynomials for $\alpha = \beta = \lambda - 1/2$, $(-\frac{1}{2} < \lambda < \frac{1}{2})$ and is denoted by $P_n^{(\lambda)}(x)$. For $\alpha = \beta = 0$, it is the Legendre polynomials $P_n(x)$. For $\alpha = \beta = -\frac{1}{2}$ and $\alpha = \beta = +\frac{1}{2}$, Jacobi polynomials are *Tchebichef* polynomials of *first* and *second* kinds respectively.

1.2 The kernel For many useful purposes in analysis, following sums called "kernel" are used as important tools (Szegö[1] p. 71).

(1.2.1)
$$K_n^{(\alpha,\beta)}(x,y) = \sum_{v=0}^{n} \left\{ h_v^{(\alpha,\beta)} \right\} P_v^{(\alpha,\ \beta)}(x) P_v^{(\alpha,\ \beta)}(y)$$

$$= \frac{2^{-\alpha-\beta}}{2n+\alpha+\beta+2} \frac{\Gamma(n+2)\,\Gamma(n+\alpha+\beta+2)}{\Gamma(n+\alpha+1)\,\Gamma(n+\beta+1)}$$

$$\frac{P_{n+1}^{(\alpha,\ \beta)}(x)\ P_n^{(\alpha,\ \beta)}(y) \quad - \quad P_n^{(\alpha,\ \beta)}(x)\,P_{n+1}^{(\alpha,\ \beta)}(y)}{x-y}$$

For y = 1,

(1.2.2)
$$K_n^{(\alpha,\beta)}(x,y) \equiv K_n^{(\alpha,\beta)}(x)$$

$$= \frac{2^{-\alpha-\beta-1}}{\Gamma(\alpha+1)\,\Gamma(n+\beta+1)}\ \frac{\Gamma(n+\alpha+\beta+2)}{}\ P_n^{(\alpha+1,\ \beta)}(x)$$

If we divide $P_n^{(\alpha,\ \beta)}(x)$ by its end point value at $x = 1$, we get another look of Jacobi polynomials denoted with meaning

(1.2.3)
$$R_n^{(\alpha,\ \beta)}(x) \equiv \frac{P_n^{(\alpha,\ \beta)}(x)}{P_n^{(\alpha,\ \beta)}(1)} = \frac{\Gamma(n+1)\,\Gamma(\alpha+1)}{\Gamma(n+\alpha+1)}\ P_n^{(\alpha,\ \beta)}(x)$$

Convergence and approximation problems are also studied with this notations as at $x = 1$

(1.2.4)
$$R_n^{(\alpha,\ \beta)}(1) = 1$$

$R_n^{(\alpha,\ \beta)}(x)$ is called **normal** form of Jacobi polynomials. Rewriting of above formulae is easy. (1.1.7) has the form

(1.2.5)
$$\int_{-1}^{+1}(1-x)^{\alpha}(1+x)^{\beta}\ \left\{ R_n^{(\alpha,\beta)}(x) \right\}^2\ dx$$

$$= \frac{2^{\alpha+\beta+1}\{\Gamma(\alpha+1)\}^2}{2n+\alpha+\beta+1}\ \frac{\Gamma(n+1)\,\Gamma(n+\beta+1)}{\Gamma(n+\alpha+1)(\Gamma n+\alpha+\beta+1)}$$

$$= \left(\omega_n^{(\alpha,\beta)} \right)^{-1} \quad \text{(say)},$$

and the form of 'the *Kernel*" (1.2.2) is

$$K_n^{(\alpha,\beta)}(x) \equiv \sum_{v=0}^{n} \{h_v^{(\alpha,\beta)}\}^{-1} \, P_v^{(\alpha,\beta)}(x) \, P_v^{(\alpha,\beta)}(1)$$

$$= 2^{-\alpha-\beta-1} \frac{\Gamma(n+\alpha+\beta+2)}{\Gamma(\alpha+1)\Gamma(n+\beta+1)} \, P_n^{(\alpha+1,\beta)}(x)$$

$$= 2^{-\alpha-\beta-1} \frac{\Gamma(n+\alpha+\beta+2)}{\Gamma(\alpha+1)\Gamma(n+\beta+1)} \, P_n^{(\alpha+1,\beta)}(1) \, R_n^{(\alpha+1,\beta)}(x)$$

(see Szego[1](4.3.5))

$$= 2^{-\alpha-\beta-1} \frac{\Gamma(n+\alpha+\beta+2)\Gamma(n+\alpha+2)}{\Gamma(\alpha+1)\Gamma(n+\beta+1)\Gamma(n+1)\Gamma(\alpha+2)} \, R_n^{(\alpha+1,\beta)}(x)$$

(1.2.6)
$$= \frac{2(\alpha+1)}{2n+\alpha+\beta+2} \, \omega_n^{(\alpha+1,\beta)} \, R_n^{(\alpha+1,\beta)}(x) \qquad \text{(by (1.2.3))}$$

where $\omega_n^{(\alpha+1,\beta)}$ is Jacobi number defined in (2.1.7) given by (1.2.5).

1.3 Inequalities. We are presenting some basic inequalities from Szegö[2]. Whereas we will derive the necessary orders of the Jacobi polynomials when and where ever which one is applicable. For α, β arbitrary real and $n \to \infty$,

(1.3.1a)
$$P_n^{(\alpha,\beta)}(\cos\vartheta) = \begin{cases} \vartheta^{-\alpha-\frac{1}{2}} O\left(n^{-\frac{1}{2}}\right), & \text{if } cn^{-1} \leq \vartheta \leq \frac{\pi}{2} \\ O(n^\alpha), & \text{if } 0 \leq \vartheta \leq cn^{-1}, \end{cases}$$

(see Theorem 7.32.2 of Szegö[1]) and in $0 < \vartheta \leq \frac{\pi}{2}$, for all $\alpha \geq -1/2$

(1.3.1b)
$$P_n^{(\alpha,\beta)}(\cos\vartheta) = \begin{cases} \vartheta^{-\alpha-\frac{1}{2}} O\left(n^{-\frac{1}{2}}\right) \\ O(n^\alpha), \end{cases}$$

Again in $0 < \vartheta \leq \frac{\pi}{2}$, for all $\alpha \leq -1/2$

(1.3.2)
$$P_n^{(\alpha,\beta)}(\cos\vartheta) = O(n^\alpha).$$

Similar results are drawn in $(\frac{\pi}{2},\pi)$ by symmetry in view of (1.1.4). For $\alpha, \beta > -1$, one has

(1.3.3)
$$max_{-1 \leq x \leq +1} \left| P_n^{(\alpha,\beta)}(x) \right|$$
$$= \begin{cases} \binom{n+q}{n} \sim n^q & \text{if } q = max(\alpha,\beta) \geq -1/2, \\ O\left(n^{-\frac{1}{2}}\right) & \text{if } q = max(\alpha,\beta) < -1/2, \end{cases}$$

For an integral bound, we have

$$(1.3.4) \quad \int_0^1 (1-x)^\mu \left| P_n^{(\alpha,\beta)}(x) \right| dx \ \sim \ \begin{cases} n^{\alpha-2\mu-2}, & 2\mu \ < \ \alpha - 3/2 \\ n^{-\frac{1}{2}} \log n, & 2\mu \ = \ \alpha - 3/2 \\ n^{-\frac{1}{2}}, & 2\mu \ > \ \alpha - 3/2 \end{cases}$$

when $\alpha, \beta, \mu > -1$. For arbitrary α, β, in $0 < \vartheta < \pi$,

$$(1.3.5) \quad P_n^{(\alpha,\beta)}(\cos\vartheta) \ = \ n^{-\frac{1}{2}} k(\vartheta) \, \cos(N\vartheta + \gamma) + O\left(n^{-\frac{3}{2}}\right).$$

where

$$(1.3.6) \quad k(\vartheta) \ = \ \pi^{-1/2}(\sin\vartheta)^{-\alpha-\frac{1}{2}}(\cos\vartheta)^{-\beta-\frac{1}{2}}, \quad N = n + \frac{\alpha+\beta+1}{2},$$

$$\gamma \ = \ -\left(\alpha + \frac{1}{2}\right)\pi/2, \quad \text{(see Darboux [1])}.$$

The bound in (1.3.5) holds uniformly in $\varepsilon \le \theta \le \pi - \varepsilon$ for every $\varepsilon > 0$. Also in $cn^{-1} \le \vartheta \le \pi - cn^{-1}$ for c fixed positive number

$$(1.3.7) \quad P_n^{(\alpha,\beta)}(\cos\vartheta) \ = \ n^{-\frac{1}{2}} k(\vartheta) \, \{ \cos(N\vartheta + \gamma) + (n\sin\vartheta)^{-1} O(1).$$

is due to Szegö [2] proved for $\alpha, \beta > -1$. With this preparation we proceed to frame JACOBI series, some times called Fourier–Jacobi series or in short F-J series in the next chapter.

CHAPTER II

JACOBI SERIES: EXISTENCE AND CONVERGENCE

2.1 Orthogonal expansion

Let us denote by X the spaces C or $X_p^{\alpha,\beta}$, $(\alpha, \beta > -1; 1 \le p \le \infty)$. By C we mean space of all continuous functions on $[-1, +1]$ and by $X_p^{\alpha,\beta}$ the space of p-power Lebesgue integrable functions on $[-1, +1]$ with weight $w(x) = (1 - x)^{\alpha}(1 + x)^{\beta}$. Moreover, for expansion of $f \in X$ in a Jacobi series, it is mandatory that every element of X is required to be Lebesgue integrable with weight w (see Szegö[1] discussion above eqn. (9.11.1)). It is clear that X is a Banach space with *sup, p* or *ess sup* norms. An infinite series

(2.1.1) $\quad a_0 P_0^{(\alpha,\beta)}(x) + a_1 P_1^{(\alpha,\beta)}(x) + a_2 P_2^{(\alpha,\beta)}(x) + \dots + a_n P_n^{(\alpha,\beta)}(x)$

$$\equiv \sum_{k=0}^{\infty} a_k P_k^{(\alpha,\beta)}(x)$$

where a_i (i= 0, 1, 2, ...) are real or complex numbers, is a series of Jacobi polynomials. If the coefficients a_k are defined by

(2.1.2a) $\qquad h_k^{(\alpha,\beta)} a_k = \int_{-1}^{1} f(x) \, P_k^{(\alpha,\beta)}(x) \, w(x) \, dx$

where

(2.1.2b) $\qquad h_k^{(\alpha,\beta)} = \int_{-1}^{1} \left\{ P_k^{(\alpha,\beta)} \right\}^2 w(x) \, dx$

$$= \frac{2^{\alpha+\beta+1}}{2k+\alpha+\beta+1} \frac{\Gamma(k+\alpha+1)\Gamma(k+\beta+1)}{\Gamma(k+1)\Gamma(k+\alpha+\beta+1)}$$

(see Szegö [1] (4.3.3)). Then the series (2.1.1) is called Jacobi series or Fourier-Jacobi expansion or in short F-J series associated with the function $f(x)$ from which the coefficients are drawn. If the function $f(x)$ is Lebesgue integrable with weight $w(x)$ i.e. if

(2.1.2c) $\qquad \int_{-1}^{1} |f(x)| \, w(x) \, dx < \infty$

then every integral in (2.1.2a) will exist because of the absolute continuity of Jacobi polynomials. In this way we consider a space $X \equiv X_p^{(\alpha,\beta)} \cap L_1(w)$ whenever we speak of a Jacobi series. Thus the Jacobi series associated with an $f(x)$ will exist provided that the function f is Lebesgue integrable with weight w. After the existence another step in this direction is to consider the convergence problem of the Jacobi series

(2.1.3) $\qquad f(x) \sim \sum_{k=0}^{\infty} a_k P_k^{(\alpha,\beta)}(x)$

which means a Jacobi series associated with $f \in X \equiv X_p^{(\alpha, \beta)} \cap L_1(w)$. If we use the normal form of Jacobi polynomials $R_n^{(\alpha, \beta)}(x)$ the series (2.1.3) takes the form

(2.1.4) $\qquad f(x) \sim \sum_{k=0}^{\infty} a_k P_k^{(\alpha, \beta)}(x) \equiv \sum_{k=0}^{\infty} \hat{f}(k) \, \omega_k^{(\alpha, \beta)} R_k^{(\alpha, \beta)}(x)$

where $\hat{f}(k)$ is Fourier-Jacobi transform given by

(2.1.5) $\qquad \hat{f}(k) = \int_{-1}^{1} f(t) \, R_k^{(\alpha, \beta)}(t) \, w(t) \, dt$

and $\omega_k^{(\alpha, \beta)}$ called Jacobi numbers are given by

(2.1.6) $\qquad \omega_k^{(\alpha, \beta)} = \left(\int_{-1}^{+1} \left\{ R_k^{(\alpha, \beta)}(t) \right\}^2 w(x) \, dx \right)^{-1}$

$$= \frac{(2k + \alpha + \beta + 1)}{2^{\alpha + \beta + 1} \{\Gamma(\alpha + 1)\}^2} \frac{\Gamma(k + \alpha + \beta + 1)\Gamma(k + \alpha + 1)}{\Gamma(k + 1)\Gamma(k + \beta + 1)} \quad (k = 0, 1, 2, 3, \ldots).$$

Value of $\omega_k^{(\alpha, \beta)}$ in (2.1.6) is asymptotically

(2.1.7) $\qquad = k^{2\alpha + 1}(1 + O(1/k)) = k^{2\alpha + 1} L(k) \quad \text{(say)},$

where $L(k) = (1 + O\left(\frac{1}{k}\right))$ is bounded and slowly varying function of k such that $\{k^{-\epsilon} L(k)\}$, $(\epsilon > 0)$ is decreasing from certain $k > k_1$ (fixed) and $\{k^{\delta} L(k)\}$, $(\delta > 0)$ is increasing from fixed k_2 when $k > k_2$. Since for every value of ϑ variable $\cos \vartheta$ lies in $[1, +1]$. So using $\cos \vartheta$ in place of x one has the form of (2.1.4) as

(2.1.8) $\qquad f(\cos \vartheta) \sim \sum_{k=0}^{\infty} a_k P_k^{(\alpha, \beta)}(\cos \vartheta) \equiv \sum_{k=0}^{\infty} \hat{f}(k) \, \omega_k^{(\alpha, \beta)} \, R_k^{(\alpha, \beta)}(\cos \vartheta)$

Without further reference we will use the format (2.1.8) in place of (2.1.4).

2.2 Convergence of Jacobi series

Jacobi series is a well read subject. Some specific names in the field are Haar [1], W. H. Young[1], Admoff[1], Kogbetliantz[1] [2], Szegö [1], [2], Gupta [1],[2],[3],[4], Pandey[1] and Yadav[1-23]. Particular cases of Jacobi series at the poles ± 1 and at the interior points of the interval [- 1, +1] are studied by Lukacs[1], Hilb[1] and Fezér[1, 2]. We are heavily dependent on the treatise of Szegö [1] 'The Orthogonal Polynomials' which includes all these historically important facts. We believe that our treatment of the subject sets Jacobi series on a sound footing. Following result by F. Newman given in Whitakar and Watson[1] is basic and oldest one.

THEOREM 2.2.1 (Expansion of an analytic function in a Jacobi series)

Let f(x) be analytic on the closed segment [- 1, +1]. The expansion of f(x) in a Jacobi series is convergent in the interior of the greatest ellipse with foci at ± 1 in which f(x) is regular. The

expansion is divergent in the exterior of this ellipse. Let R be the sum of semi-axes of the ellipse of convergence, then

(2.2.1) $$R \;=\; \lim_{n \to \infty} \inf |a_n|^{-1/2}$$

where a_n is coefficient of nth term of the expansion (2.1.1).

REMARK 2.2.2   Szegö ([1] see Theorem 9.2.1) has referred Whittakar and Watson [1] for the details where a shortened proof of this theorem has been given. A simplified discussion of the proof of above theorem is being presented below.

Proof of Theorem 2.2.1   For regular function $f(x)$ the expansion

(2.2.2) $$\sum_{n=0}^{\infty} \left\{ h_n^{\alpha,\beta} \right\}^{-1} P_n^{(\alpha,\beta)}(x)\, Q_n^{(\alpha,\beta)}(y) \;=\; \tfrac{1}{2} \frac{(y-1)^{-\alpha}(y+1)^{-\beta}}{y-x}$$

holds, where $Q_n^{(\alpha,\beta)}(x)$ is Jacobi polynomial of second kind (see Szegö [1] sec. § 4.61 and (4.3.3)). While x lies in the interior of and y in the exterior of an arbitrary ellipse with foci at ± 1. From the theory of analytic functions, it follows that

$$(y-1)^{\alpha}(y+1)^{\beta} Q_n^{(\alpha,\beta)}(y)$$

is single valued and regular in the complex y-plane cut along [-1, +1].

Let $f(x)$ be regular when x is in the interior of the ellipse enclosing $|z| = R > 1$. We multiply (2.2.2) by

$$(\pi i)^{-1}(y-1)^{\alpha}(y+1)^{\beta} f(y)$$

and integrate over the ellipse enclosing $|\zeta| = R - \varepsilon,\; 0 < \varepsilon < R/2$ to get

(2.2.3) $$\frac{1}{2\pi i} \int_{|\zeta|} \frac{f(y)}{y-x}\, dy \;=$$

$$\frac{1}{\pi i} \int_{R-\varepsilon} \sum_{n=0}^{\infty} \left\{ h_n^{\alpha,\beta} \right\}^{-1} (y-1)^{\alpha}(y+1)^{\beta} P_n^{(\alpha,\beta)}(x)\, Q_n^{(\alpha,\beta)}(y)\, dy$$

which is equivalent to

(2.2.4) $$f(x) \;=\; \sum_{n=0}^{\infty} a_n P_n^{(\alpha,\beta)}(x)$$

by Cauchy integral formula, for

(2.2.5) $$a_n \;=\; \left\{ \pi i \left\{ h_n^{(\alpha,\beta)} \right\} \right\}^{-1} \int_{R-\varepsilon} (y-1)^{\alpha}(y+1)^{\beta} Q_n^{(\alpha,\beta)}(y)\, dy$$

On the line segment [-1, +1], the series in (2.2.4) gives Fourier-Jacobi expansion of an analytic function $f(x)$ and the series converges uniformly to $f(x)$. Moreover, on multiplying both the sides of (2.2.4) by $P_n^{(\alpha,\beta)}(x)$ and integrating over [-1, +1] with weight $w(x) = (1-x)^{\alpha}(1+x)^{\beta}$, we get

(2.2.6) $\qquad \int_{-1}^{1} f(x) \ P_n^{(\alpha,\ \beta)}(x) \ w(x) \ dx \ = \ a_n \ \int_{-1}^{1} f(x) \ \left\{ P_n^{(\alpha,\ \beta)}(x) \right\}^2 w(x) \ dx$

$$= \ a_n \ h_n^{(\alpha,\ \beta)}$$

where other terms on the right become zero for the degree of polynomials being other than n. This shows that a_n is well defined. Thus conclusion of the theorem 2.1.1 follows from (2.2.4) and (2.2.6) while value of R is a consequence of Cauchy-Hadmard formula. This completes the proof of Theorem 2.1.1.

We mention the important equiconvergence theorem of Szegö[1] and give some clarification in finishing the proof of the theorem. Its importance lies in applying the similar results of Fourier series to Jacobi series at an interior point of [-1, +1]. At the same times it is noted that convergence of Jacobi series at the poles ± 1 is quite different.

THEOREM 2.2.3 **(Equiconvergence theorem for Jacobi series in the interior of**

the interval [-1, +1])

Let f(x) be Lebesgue measurable in [-1, +1] and let the integrals

(2.2.7) $\qquad \int_{-1}^{+1} (1 - x)^\alpha (1 + x)^\beta |f(x)| \ dx$

and

$$\int_{-1}^{+1} (1 - x)^{\frac{\alpha}{2} - \frac{1}{4}} (1 + x)^{\frac{\beta}{2} - \frac{1}{4}} \ |f(x)| \ dx$$

exist. If $s_n(x)$ denotes the nth partial sum of the expansion of f(x) in a Jacobi series and $S_n(\cos \theta)$ the nth partial sum of the cosine Fourier series of

(2.2.8) $\qquad \left(1 - \cos \theta\right)^{\frac{\alpha}{2} + \frac{1}{4}} \left(1 + \cos \theta\right)^{\frac{\beta}{2} + \frac{1}{4}} f(\cos \theta)$

Then for -1 < x < +1,

(2.2.9) $\qquad lim_{n \to \infty} \left\{ s_n(x) - (1 - x)^{-\frac{\alpha}{2} - \frac{1}{4}} (1 + x)^{-\frac{\beta}{2} - \frac{1}{4}} S_n(x) \right\} = 0$

uniformly in - 1 + ε ≤ x ≤ 1 - ε, where ε is a fixed positive number less than 1.

Proof. The difference (2.2.9) admit the presentation

(2.2.10) $\qquad \int_{-1}^{+1} f(t) \ \bigg[(1 - t)^\alpha (1 + t)^\beta \ K_n^{(\alpha,\ \beta)}(x, t) -$

$$(1 - x)^{-\frac{\alpha}{2} - \frac{1}{4}} (1 + x)^{-\frac{\beta}{2} - \frac{1}{4}} \ (1 - t)^{\frac{\alpha}{2} - \frac{1}{4}} (1 + t)^{\frac{\beta}{2} - \frac{1}{4}} \ K_n^{\left(-\frac{1}{2},\ -\frac{1}{2}\right)}(x, t) \bigg] \ dt$$

See (Szegö[1] (9.11.4)) where $K_n^{(\alpha,\ \beta)}(x,t)$ is given and $P_n^{\left(-\frac{1}{2},\ -\frac{1}{2}\right)}(x) = A\,cos\,nx$ with A, some constant is used. If $f(x) \equiv \rho(x)$ a polynomial of degree less than n then $s_n(x) = \rho(x)$ and

$$(1-x)^{-\frac{\alpha}{2}-\frac{1}{4}}(1+x)^{-\frac{\beta}{2}-\frac{1}{4}}S_n(x) \rightarrow \rho(x) \quad \text{as, } n \rightarrow \infty.$$

By a proper choice of $\rho(x)$ the integral

(2.2.11) $\qquad \int_{-1}^{+1}(1-t)^a(1+t)^b \big| f(t) - \rho(t)\big|\,dt,$

where

$$\begin{cases} a &= min\,(\alpha,\ \frac{\alpha}{2}-\frac{1}{4}) \\ b &= min(\beta,\ \frac{\beta}{2}-\frac{1}{4}) \end{cases}$$

is arbitrarily small. Thus the difference (2.2.10) must at least admit the estimate

(2.2.12) $\qquad O(1)\int_{-1}^{+1}(1-t)^a(1+t)^b|f(t)|\,dt + o(1)$

where both bounds O(1) and o(1) hold uniformly in x when $-1+\varepsilon \leq x \leq 1-\varepsilon$. But the first part of the difference (2.2.10) after calculation is

(2.2.13) $\qquad \frac{1}{2\pi}\int_\eta^{\pi-\eta} f(cos\ \varphi)\dfrac{sin(2n+1)\frac{\varphi-\vartheta}{2}}{sin\frac{\varphi-\vartheta}{2}}\,d\varphi + O(1)\int_{-1+\frac{\varepsilon}{2}}^{1-\frac{\varepsilon}{2}}|f(t)|\,dt + o(1)$

as $n \rightarrow \infty$ and second of that is

(2.2.14) $\qquad \frac{1}{2\pi}\left(1-cos\,\theta\right)^{\frac{\alpha}{2}+\frac{1}{4}}\left(1+cos\,\theta\right)^{\frac{\beta}{2}+\frac{1}{4}}\int_\eta^{\pi-\eta} f(cos\ \varphi)\dfrac{sin(2n+1)\frac{\varphi-\vartheta}{2}}{sin\frac{\varphi-\vartheta}{2}}\,d\varphi$

$$+\ O(1)\int_{-1+\frac{\varepsilon}{2}}^{1-\frac{\varepsilon}{2}}|f(t)|\,dt + o(1)\ .$$

For detail calculations, a helpful consultation of (Szegö[1] eqns. (9.3.6) and (9.3.7)) is suggestive. The estimate (2.2.12) holds in $\left[1-\frac{\varepsilon}{2},\ 1\right]$ and $\left[-1+\frac{\varepsilon}{2},\ -1\right]$. Thus the conclusion (2.2.9) is established. This is sufficient to conclude the result of the Theorem 2.2.3 This may be treated as a clarification of what has been done in Szegö[1]. In the last it is to be cleared that both the integrals in (2.2.7) are required. To see this, we observe the general term of Jacobi series of a function

(2.2.15) $\qquad f(x) = (1-x)^\mu,\quad -1-\alpha\ <\mu\ \leq -\frac{\alpha}{2}-\frac{3}{4}\ ;$

which is asymptotically equal to

(2.2.16) $$n^{-2\mu-\alpha-\frac{3}{2}} \cos(N\vartheta + \gamma), \quad (\mu \neq 0,1,2,\ldots)$$

in the interior of $[-1, +1]$ (see Szegö[1] (9.3.11), N and γ as given in (1.3.6)). If we take $\alpha > -\frac{1}{2}$, $0 < \vartheta < \pi$ the Jacobi series whose principal term is given by (2.2.16) diverges for

(2.2.17) $$-1 - \alpha < \mu \leq -\frac{\alpha}{2} - \frac{3}{4}$$

because by (2.217), $-2\mu - \alpha - \frac{3}{2} \geq 0$. For the range $-1 - \alpha < \mu$ *i.e.* $(\mu + \alpha > -1)$ the first integral in (2.2.7) exists but not the second. As then

(2.2.18) $$\frac{\alpha}{2} - \frac{1}{4} + \mu \leq \frac{\alpha}{2} - \frac{1}{4} - \frac{\alpha}{2} - \frac{3}{4} \leq -1$$

which shows nonexistence of second integral in (2.2.7) and gives divergent Jacobi series, making the Theorem meaningless. This suggests that to see the truth of the Theorem, existence of both the integrals in (2.2.7) are required. This completes the proof of Theorem 2.2.3.

2.3   Convergence by an application of summability method

Following summability theorems given in Szegö[1] and well-known in the literature are like platforms to study the Jacobi series and related problems. To make the subject interesting, we survey the same in some graspable way.

THEOREM   2.3.1   (Summability theorem at end points)

Let f(x) be continuous on the closed segment [-1, +1]. The expansion of f(x) in a Jacobi series is (C, k)-summable at x = +1, provided k > α + ½. This is in general not true if k = α + ½. An analogous statement holds for x = − 1, α being replaced by β.

Proof.   A summarized proof of this theorem is being presented for the sake of completeness. Since f is continuous, consequently by Helly's theorem (Theo. 1.6 in Szegö[1]) we have to prove the boundedness of the sequence

(2.3.1) $$L_n^{(k)} = \left\{C_n^{(k)}\right\}^{-1} \int_{-1}^{+1} \left|S_n^{(k)}(x)\right| w(x)\, dx$$

if and only if $k > \alpha + \frac{1}{2}$ where $L_n^{(k)}$ is Lebesgue constant of nth Cesàro mean of order k of (2.1.1). $S_n^{(k)}(x)$ is numerator of Cesàro mean of order k of the F-J series. If we denote by $\sigma_n^{(k)}(x)$ the nth Cesàro of order k of (2.1.1). Then at $x = 1$,

(2.3.2)
$$\sigma_n^{(k)}(1) = \left\{C_n^{(k)}\right\}^{-1} \int_{-1}^{+1} f(x) \sum_{m=0}^{n} C_{n-m}^{(k)} \left\{h_m^{(\alpha,\beta)}\right\}^{-1} P_m^{(\alpha,\beta)}(1) P_m^{(\alpha,\beta)}(x)\, w(x)\, dx$$

(see the notations (1.1.7)). Thus the Lebesgue constants are given by

(2.2.3)
$$L_n^{(k)} = \left\{ C_n^{(k)} \right\}^{-1} \int_{-1}^{+1} \left| \sum_{m=0}^{n} C_{n-m}^{(k)} \left\{ h_m^{(\alpha, \beta)} \right\}^{-1} \right.$$

$$\left. P_m^{(\alpha, \beta)}(1) \, P_m^{(\alpha, \beta)}(x) \right| \, w(x) \, dx$$

which on calculation is

(2.2.4)
$$= O(n^{-k}) \sum_{v=0}^{n} \left| G_v(n, k) \right| v^{2\alpha + 1}$$

(see Szegö[1] (9.41.3)) where

(2.3.5)
$$G_v(n, k) = \sum_{\rho=0}^{k-1} (n - v)^\rho O(v^{-k-\rho-2})$$

But

$$L_n^{(k)} = O(n^{-k}) \sum_{\rho=0}^{k-1} n^\rho + O(n^{-k}) \sum_{v=1}^{n-1} \sum_{\rho=0}^{k-1} (n-v)^\rho (v^{-k-\rho-2}) v^{2k+1}$$

$$+ O\left(n^{-k}\right) \left\{ \sum_{\rho=0}^{k-1} n^{-k-\rho-2} n^{2k+1} + n^{-k-1} n^{2k+1} \right\}$$

(2.3.6) $\quad = O(1)$ \qquad\qquad (for k non-negative integer).

Let $k > \alpha + \frac{1}{2}$., k not an integer. Then for $k' = [k] + 1$

$$L_n^{(k')} \le A \quad \text{(by the same argument as above)}.$$

So by the regularity we have

$$L_n^{(\sigma)} \le A \qquad \text{(for } \sigma \ge k', \; n = 0, 1, 2, 3, \ldots).$$

Thus for k not an integer and $k > \alpha + \frac{1}{2}$,

(2.3.7) \qquad $L_n^{(k)} \le A$,

(for details see Szegö[1] (9.41.16)). Again for $k = \alpha + \frac{1}{2}$, it is argued that

(2.3.8) \qquad $L_n^{(k)} > A \, \log n$,

which shows that for a continuous function *f(x)* a Fourier–Jacobi expansion (2.2.1) is not (C, k)-summable. Thus by regularity, it is finally concluded that the Jacobi series associated with a continuous function is not (C, k)-summable for $k \leq \alpha + \frac{1}{2}$. This completes the proof of the Theorem 2.3.1.

REMARK 2.3.3 This theorem has an extensive literature (Szegö[1] § 9.11). F-J expansion is (C, k)-summable for $\alpha > -1$, $k > \alpha + \frac{1}{2}$. This in turn implies the convergence of F-J series (case $k = 0$) at the pole $x = +1$. Also a side by side decision is that F-J series of a continuous function converges for $-1 < \alpha < -\frac{1}{2}$ but may diverge for $\alpha \geq -\frac{1}{2}$ at the pole $x = +1$ and similar decision holds at the other pole $x = -1$ only change in α, β is apparent.

THEOREM 2.3.3 (**Generalized summability theorem 9.1.4 of Szegö**)

Let f(x) be Lebesgue measurable in [- 1, +1] and continuous at $x = +1$. Then if we assume the existence of the integral

(2.3.9) $$\int_{-1}^{+1}(1 - x)^{\alpha}(1 + x)^{\beta}|f(x)|\, dx,$$

the Jacobi series is (C, k)-summable $k > \alpha + \frac{1}{2}$, at $x = +1$, provided that in the case

(2.3.10) $$\beta > -\frac{1}{2}, \quad \alpha + \frac{1}{2} < k < \alpha + \beta + 1,$$

the following additional 'antipole' condition is satisfied: the integral

(2.3.11) $$\int_{-1}^{0}(1 + x)^{\frac{\beta}{2} - \frac{1}{4}} |f(x)|\, dx,$$

exists. (For $k \geq \alpha + \beta + 1$ no antipole condition is necessary) For $k \leq \alpha + \frac{1}{2}$ or for $k > \alpha + \frac{1}{2}$ but without the antipole condition, the statement is not true.

Proof of Theorem 2.3.3 is thoroughly explained in Szegö [1] chapter IX and is based on the following lemma.

LEMMA 2.3.4 *If f(x) is continuous at $x = 1$ and $k > \alpha + \frac{1}{2}$. Then under the conditions of the theorem 2.3.3, we have*

(2.3.12) $$\int_{-1}^{+1}|f(x) - f(1)|\, P_n^{(\alpha + k + 1,\ \beta)}(x)\, w(x)dx = o(n^{k - \alpha - 1})$$

Proof. For ε arbitrarily small positive number, let us break the interval of integration in (2.3.12) into $0 \leq \theta \leq \varepsilon$, $\varepsilon \leq \theta \leq \pi - \varepsilon$ and $\pi - \varepsilon \leq \theta \leq \pi$ after substituting $x = \cos\theta$ denote each one by T_1, T_2 and T_3 respectively. Thus for $k > \alpha + \frac{1}{2}$,

$$T_1 = max_{0 \leq \theta \leq \varepsilon} |f(x) - f(1)| \int_{\cos\varepsilon}^{+1} \left| P_n^{(\alpha + k + 1,\ \beta)}(x) \right| w(x)dx$$

$$= o(1)\, O(n^{k - \alpha - 1}) \qquad \text{(by (1.3.4))},$$

$$T_2 = O(n^{-\frac{1}{2}}) = o(n^{k - \alpha - 1}), \quad \text{(by (1.3.5))}$$

and

$$T_3 = \begin{cases} \int_{\pi - \varepsilon}^{\pi} |f(\cos\theta) - f(1)| (\pi - \theta)^{2\beta + 1} O(n^{-\frac{1}{2}}) \, d\theta = o(n^{k - \alpha - 1}) \\ \qquad (for -1 < \beta \le -\frac{1}{2} \ by \ (1.3.2)) \\ \int_{\pi - \varepsilon}^{\pi} |f(\cos\theta) - f(1)| (\pi - \theta)^{2\beta + 1} O(n^{\beta}) \, d\theta = o(n^{\beta}) = o(n^{k - \alpha - 1}) \\ \qquad (for \ \varepsilon \ arbitrary \ -\frac{1}{2} < \beta \le k - \alpha - 1) \\ \int_{\pi - \varepsilon}^{\pi} |f(\cos\theta) - f(1)| (\pi - \theta)^{2\beta + 1} (\pi - \theta)^{-\beta - \frac{1}{2}} O(n^{-\frac{1}{2}}) \, d\theta = o(n^{k - \alpha - 1}) \\ \qquad (for \ \beta > k - \alpha - 1, by \ 'antipole' \ condition.) \end{cases}$$

This completes the proof of the lemma.

Proof of Theorem 2.3.3   Let us denote by $\sigma_n^{(k)}(x)$ the nth Cesàro mean of order k of F-J series (2.1.1). Then using the notations of Szegö[1] at $x = 1$

$$\left| \sigma_n^{(k)}(1) - f(x) \right| \le \left\{ C_n^{(k)} \right\}^{-1} \int_{-1}^{+1} |f(x) - f(1)| \left| S_n^{(k)}(x) \right| w(x) dx$$

(2.3.13)
$$= M_n^{(k)} \quad (\text{say}),$$

where $S_n^{(k)}(x)$ is numerator of the Cesàro mean. For positive integer k,

$$M_n^{(k)} = O\left(n^{-k}\right) \sum_{\rho = 0}^{k-1} n^{\rho} + O\left(n^{-k}\right) \sum_{\nu = 1}^{n-1} \sum_{\rho = 0}^{k-1} (n - \nu)^{\rho} (\nu^{-k - \rho - 2}) o(\nu^{2k + 1})$$

$$+ O\left(n^{-k}\right) \left\{ \sum_{\rho = 0}^{k-1} n^{-k - \rho - 2} o(n^{2k + 1}) + n^{-k-1} o(n^{2k+1}) \right\}$$

(2.3.14)
$$= o(1) \quad (\text{see Szegö[1] (9.42.4)}).$$

Argument of Szegö[1] (9.42.6) works for all $k > \alpha + \frac{1}{2}$ and the case $k \le \alpha + \frac{1}{2}$ is treated as argued in (2.3.8). We believe that our survey of proof of Theorem 2.3.1 is complete.

REMARK 2.3.5   Continuity of $f(x)$ at $x = 1$ can be replaced by a more general condition

(2.3.15)
$$\int_0^t |f(\cos\varphi) - f(1)| \, d\varphi = o(t) \quad \text{as } t \to +0.$$

Calculation of T_1 above in the proof of lemma 2.3.4 is altered by breaking the interval of integration $0 \le \theta \le \varepsilon$ into $0 \le \theta \le c n^{-1}$ and $c n^{-1} \le \theta \le \varepsilon$. Then one uses the valid orders of (1.3.1).

REMARK 2.3.6   By an example it is demonstrated that the Jacobi series is not (C, k)-summable if the antipole condition is not satisfied. Let

(2.3.16) $f(x) = (1 + x)^\mu, \quad -1-\beta < \mu \le \frac{1}{2}(\alpha - \beta - k - 1)$

Its principal term in the expansion of Jacobi series is

(2.3.17) $(-1)^n n^{\alpha - \beta - 2\mu - 1}$ or $(-1)^n C_n^{(\alpha - \beta - 2\mu - 1)}$

(Szegö[1] (9.42.10)). We note that the expansion is not (C, k)-summable for

(2.3.18) $k \le \alpha - \beta - 2\mu - 1 = \lambda$ (say),.

But for $(-1 - \beta < \mu)$ the expansion holds and Jacobi series exists whence it is not (C, k)-summable for even $k > \alpha + \frac{1}{2}$.

With these remarks given as a part of the proof of Theorem 2.3.3, the required proof is complete. This Theorem draws a boundary line beyond which the F-J expansion is not (C, k)-summable for $k > \alpha + \frac{1}{2}$. A noteworthy generalization of 2.3.3 has been proved by Yadav[13] which covers the functions beyond the range of Theorem 2.3.3. For the vivid clarifications we rewrite a version of the Theorem 2.3.3 only by changing the variable. If we write $x = \cos \theta$, then the interval of integration [- 1, +1] changes to [0, π]. If, by $L_1(w)$ we mean Lebesgue integrable functions on [0, π] with weight $w(\cos \theta)$ then the form of Theorem 2.3.3 is

THEOREM 2.3.7   *If $f \in L_1(w)$ and*

(2.3.19) $\int_0^t |f(\cos \varphi) - f(1)| d\varphi = $ o(t)   *as t $\to$ +0.*

Then the Jacobi series is (C, k)- summable for $k > \alpha + \frac{1}{2}$ at $\theta = 0$, provided that in case

(2.3.20) $\beta > -\frac{1}{2}, \quad \alpha + \frac{1}{2} < k < \alpha + \beta + 1$;

the following additional 'antipole' condition is satisfied: the integral

(2.3.21) $\int_0^\varepsilon \varphi^{\beta + \frac{1}{2}} |f(-\cos \varphi)| d\varphi , \quad (\varepsilon > 0)$

exists. (For $k \ge \alpha + \beta + 1$, no antipole condition is necessary). For $k \le \alpha + \frac{1}{2}$ or, for $k > \alpha + \frac{1}{2}$ but without the antipole condition, the statement is not true.

Generalization of this Theorem by improving the '*pole*' and the '*antipole*' conditions, Yadav [13] proved the following

THEOREM 2.3.8   *If $f \in L_1(w)$ and the 'pole' condition*

(2.3.22) $\int_0^t |F(\varphi)| d\varphi = $ o(t^{2\alpha + 2})   *as t $\to$ 0+,*

where

(2.3.23) $F(\varphi) = \{f(\cos \varphi) - A\}(\sin \varphi)^{2\alpha + 1}(\cos \varphi)^{2\beta + 1}$

with A a constant depending upon f, is satisfied. Then the Jacobi series is (C, k)- summable for k > α + ½ at θ = 0, provided that in case

(2.3.24) $\beta > -½, \quad \alpha + ½ < k < \alpha + \beta + 1;$

the following additional 'antipole' condition is satisfied: the integral

(2.3.25) $\int_0^h \varphi^{\beta - \alpha + k} \left| f(-\cos \varphi) \right| d\varphi = o(1), \quad (h \to +0)$

holds. (For $-1 < \beta \le -½$ *or for* $k \ge \alpha + \beta + 1$, *no antipole condition is necessary). For* $k \le \alpha + ½$, *or for* $k > \alpha + ½$ *but without the antipole condition, the statement is not true.*

REMARK   2.3.9   The function $f(x) = \dfrac{\log\left(1 + \sqrt{\dfrac{1-x}{2}}\right)}{\sqrt{\dfrac{1-x}{2}}}$ does not satisfy the condition

(2.3.19) while it satisfies the condition (2.3.22). Again the function $(1 + x)^{-\frac{\beta}{2} - \frac{3}{4}}$, $\beta > -½$, $\alpha + ½ < k < \alpha + \beta + 1$, does not satisfy (2.3.21) while it satisfies (2.3.25). Thus the theorem is more general in the sense that it covers those functions, not in the range of Theorem 2.3.7. Details may be seen in Yadav[13]. Now the lemma 2.3.4 has the following form.

LEMMA   2.3.10   *Under conditions of Theorem 2.3.8,*

(2.3.26) $\int_0^\pi F(\varphi) \, P_n^{(\alpha + k + 1, \, \beta)}(\cos \varphi) \, d\varphi = \begin{cases} o(n^{k - \alpha - 1}), \ k > \alpha + ½ \\ o(n^{-\frac{1}{2}} \log n), \ k = \alpha + ½ \\ o(n^{-\frac{1}{2}}), \ k < \alpha + ½, \end{cases}$

as $n \to \infty$.

Proof.   The result in the lemma is estimated from five integrals made by breaking the interval of integration in (2.3.26) as

(2.3.27) $\int_0^{c/n} + \int_{c/n}^\delta + \int_\delta^{\pi - \delta'} + \int_{\pi - \delta'}^{\pi - c/n} + \int_{\pi - c/n}^\pi = \sum_{i=1}^5 T_i$ (say),

where c, δ and δ' are fixed but small positive reals. One uses (1.3.1) and (1.3.22) to estimate the orders of T_1 and T_2. Again the antipole condition is used in the estimates of T_4 and T_5. Cases $-1 < \beta \le -½$ or $k \ge \alpha + \beta + 1$ does not require the antipole condition. Using (1.3.5) and Riemman-Lebesgue lemma, we get

(2.3.28) $T_3 = O(n)^{-\frac{1}{2}} o(1) = \begin{cases} o(n^{k - \alpha - 1}), \ k > \alpha + ½ \\ o(n^{-\frac{1}{2}} \log n), \ k = \alpha + ½ \\ o(n^{-\frac{1}{2}}), \quad k < \alpha + ½. \end{cases}$

This finishes the proof of the lemma.

Proof of Theorem 2.3.8   Arguments given in the proof of Theorem 2.3.7 from (2.3.13) to (2.3.14) holds, as the results of the lemmas 2.3.4 and 2.3.10 are the same. This completes the proof of the Theorem 2.3.8.

2.4   Further results on the convergence of Jacobi series

The equiconvergence Theorem 2.1.2   settles many convergence problems of linearily transformed sequences of Jacobi series in the interior of $[0, \pi]$ as have been observed in various publications of Yadav [2], [3] and others. Convergence of Jacobi series at the end point of the interval $[-1, +1]$ is a matter of interesting logic and a serious nature of analysis involved that leads not only uniform convergence of F-J series but also relates potential results on approximations and wavelet representations. Following Theorems due to Yadav [12] and [21] are latest achievements after the convergence theorem proved by Rau [1]. We note that by X we mean the spaces given for the expansions (2.1.1) where existence of the first integral in (2.2.7) is required. Our results in (Yadav[12]) read:

THEOREM   2.4.1   *Let* $f(x) \in X$ *be continuous at* $x = 1$, *Then the Jacobi series* (2.1.8) *converges to* $f(1)$ *at* $x = 1$ *for* $-1 < \alpha < -1/2$, $-1 < \beta \leq -1/2$

(2.4.1)   *or* $(\beta > -1/2$ *but* $\alpha + \beta \leq -1$ $)$.

THEOREM   2.4.2   *Let* $f(x) \in X$ *be continuous at* $x = 1$ *and the 'antipole' condition*

(2.4.2) $\int_0^h t^\beta \left| f(-\cos t) \right| dt = o(h^\alpha)$ *as* $t \to 0+$

be satisfied. Then the Jacobi series (2.1.8) *converges to* $f(1)$ *at* $x = 1$ *for* $-1 < \alpha < -1/2$, $\beta > -1$.

THEOREM   2.4.3   *Let* $f(x) \in X$ *be continuous at* $x = 1$, *Then the Jacobi series* (2.1.8) *converges to* $f(1)$ *at* $x = 1$ *for* $-1 < \alpha < -1/2$, $\beta > -1$ *provided in case* $\beta > -1/2$ *but* $\alpha + \beta > -1$, *the additional 'antipole' condition*

(2.4.3) $\int_0^t \varphi^{\beta +1/2} \left| f(-\cos \varphi) \right| d\varphi = o(t^{\alpha + 1/2})$ *as* $t \to 0+$

is satisfied. For $-1 < \beta \leq -1/2$ *or for* $\beta > -1/2$ *but* $\alpha + \beta \leq -1$ *no antipole condition is necessary as proved in Theorem 2.4.1.*

REMARK   2.4.5   Continuity of f at $x = 1$ in the above three theorems can be replaced by a more general condition

(2.4.4a) $\int_0^t \left| f(\cos \varphi) - A \right| (\sin\varphi)^{2\alpha + 1} (\cos \varphi)^{2\beta + 1} d\varphi = o(t^{2\alpha + 2})$ as $t \to 0+$,

or in other format

(2.4.4b) $\int_0^t \left| f(\cos \varphi) - A \right| \varphi^{2\alpha + 1} d\varphi = o(t^{2\alpha + 2})$ as $t \to 0+$,

or

(2.4.4c) $\qquad \int_0^t |F(\varphi)| \, d\varphi \;=\; o(t^{2\alpha+2}) \quad$ as $\ t \to 0+$

where

(2.4.4d) $\qquad F(\varphi) \;=\; \{f(\cos\varphi) - A\}(\sin\varphi)^{2\alpha+1} (\cos\varphi)^{2\beta+1}$

constant A depends upon f. Now the F-J series shall converge to A instead of $f(1)$. The generality of (2.4.4) has been established in the Remark 2.3.9. Following lemma is key to above three Theorems.

LEMMA 2.4.6 *Under the conditions of these theorems, one has for all* $\ \beta > -1$

(2.4.5) $\qquad \int_0^\pi F(\varphi)\, P_n^{(\alpha+1,\ \beta)}(\cos\varphi)\, d\varphi \;=\; \begin{cases} o(n^{-\alpha-1}), & -1 < \alpha < -\tfrac{1}{2} \\[2mm] o(n^{-\frac{1}{2}} \log n), & \alpha = -\tfrac{1}{2} \\[2mm] o(n^{-\frac{1}{2}}), & \alpha > -\tfrac{1}{2}, \end{cases}$

as $n \to \infty$, *where* $F(\varphi)$ *is given by* (2.4.4d).

Proof. The interval of integration is broken into five parts to estimate the value of the integral.

$$\int_0^{c/n} \; + \; \int_{c/n}^{\delta} \; + \; \int_{\delta}^{\pi-\delta'} \; + \; \int_{\pi-\delta'}^{\pi-c/n} \; + \; \int_{\pi-c/n}^{\pi} \;=\; \Sigma_{i=1}^5 \, T_i \ \text{(say)},$$

where

$$T_1 \;=\; O(n^{\alpha+1}) \int_0^{c/n} |F(\varphi)| \, d\varphi \;=\; o(n^{-\alpha-1})$$

by substituting the order given in (1.3.2b) and applying the 'pole' condition (2.4.4c). Let us consider the meaning of $t \to 0+$ in the condition (2.4.4c). By this, we mean $\forall\, \varepsilon > 0, \exists\, \delta > 0 :$ $\left| \int_{c/n}^{\delta} F(\varphi)\, d\varphi \right| < \varepsilon\, \varphi^{2\alpha+2}$ whenever $c/n < \varphi < \delta$. Thus using (1.3.2b),

$$T_2 \;=\; O\left(n^{-\frac{1}{2}}\right) \int_{c/n}^{\delta} \varphi^{-\alpha-3/2} |F(\varphi)| \, d\varphi$$

$$=\; O\left(n^{-\frac{1}{2}}\right) \left\{ \left[\varphi^{-\alpha-3/2}\, \varepsilon\, \varphi^{2\alpha+2}\right]_{c/n}^{\delta} \;+\; O(1) \int_{c/n}^{\delta} \varepsilon\, \varphi^{2\alpha+2} \varphi^{-\alpha-5/2} \, d\varphi \right\}$$

$$=\; o\left(n^{-\frac{1}{2}}\right) \;+\; o\left(n^{-\alpha-1}\right) \;+\; o\left(n^{-\frac{1}{2}}\right) \int_{c/n}^{\delta} \varphi^{\alpha-1/2} \, d\varphi \quad \text{as n} \to \infty,$$

(2.4.6) $\qquad =\; \begin{cases} o(n^{-\alpha-1}), & -1 < \alpha < -\tfrac{1}{2} \\[2mm] o(n^{-\frac{1}{2}} \log n), & \alpha = -\tfrac{1}{2} \\[2mm] o(n^{-\frac{1}{2}}), & \alpha > -\tfrac{1}{2}, \end{cases} \quad \text{, as} \to \infty$

For T_3 we use (1.3.5) and apply the Remman- Lebesgue lemma to get

(2.4.7) $\qquad T_3 \;=\; O(n^{-1/2})\, o(1).$

Again with the help of the 'antipole' conditions, we get the required orders for T_4 and T_5. For detailed arguments one may see Yadav[12]. This completes the proof of the lemma.

Proof of Theorems 2.4.1, 2.4.2 *and* 2.4.3   If we denote by $s_n(f,x,X)$ the nth partial sum of (2.1.8). Then at $x = 1$, (see, Yadav [12])

$$s_n(f,1,X) \;-\; A \;=\; \int_0^\pi h_n^{(\alpha+1,\,\beta)} \, P_n^{(\alpha+1,\,\beta)}(\cos\varphi)\, P_n^{(\alpha+1,\,\beta)}(1)\, A_n\, F(\varphi)\, d\varphi$$

where $\;A_n = a_1 n^{-1} + a_2 n^{-2} + \ldots + O(n^{-\lambda})$, $(\lambda > 0)$ with a_i constants. Thus applying (2.4.5), we have

(2.4.8) $\qquad lim_{n\to\infty}\{\, s_n(f,1,X) \;-\; A\} \;\le$

$$\le\; \lim_{n\to\infty}\left[O\!\left(n^{\alpha+1}\right) \left| \int_0^\pi P_n^{(\alpha+1,\,\beta)}(\cos\varphi)\, F(\varphi)\, d\varphi \right| \; \right] = o(1).$$

This completes the proof.

It is to be noted that the case $\alpha = -\tfrac{1}{2}$ is excluded in all above Theorems while the important particular case $\alpha = \beta = -1/2$ of Jacobi polynomials is Tchebichef polynomials. It is interesting to survey the project which include this unique case. In extending the boundary of α, we were able to prove the extended forms of above Theorems in the following frame for $-1 < \alpha \le -1/2$, $\beta > -1$. But we were bound to alter the 'pole' condition (2.4.4b) generalizing the continuity at $x = 1$, as

(2.4.9) $\qquad \int_0^t \varphi^{\alpha-1/2} \left| f(\cos\varphi) - A \right| d\varphi \;=\; o(t^{\alpha+1/2})$ as $t \to 0+$.

Note that this 'pole' condition is independent of continuity or general continuity condition (2.4.4b) at $x = +1$. Moreover (2.4.9) is satisfied by the signals of the class $Lip\ \delta$, $(\delta \ge 1/4)$ in an arbitrarily small neighborhood of the poles ± 1. Also it is satisfied by those signals which have sink at the poles $x = +1$ or -1 i.e. $f(+1)$ or $f(-1) = 0$ as the case is being considered (Yadav[21]). Following Theorems are proved.

THEOREM 2.4.7   *The F-J series (2.1.8) converges to A at $x = +1$ for $-1 < \alpha \le -1/2$ and $-1 < \beta \le -1/2$ provided f satisfies (2.4.9).*

THEOREM 2.4.8   *The F-J series (2.1.8) converges to A at $x = +1$ for $-1 < \alpha \le -1/2$ and $\beta > -1/2$ with $\alpha + \beta \le -1$, provided f satisfies (2.4.9).*

THEOREM 2.4.9   *The F-J series (2.1.8) converges to A at $x = +1$ for $-1 < \alpha \le -1/2$ and $\beta > -1/2$ with $\alpha + \beta > -1$, provided f satisfies (2.4.9). and the 'antipole' condition (2.4.3).*

THEOREM 2.4.10   *The F-J series (2.1.8) converges to A at $x = +1$, for $-1 < \alpha \le -1/2$ and $\beta > -1$, provided f satisfies (2.4.9). and the 'antipole' condition (2.4.2)*

Again it is reminding that our every Theorem stated for x = 1 also holds at x = -1 only exchange of pole and antipole with α, β is apparent. Following lemma parallel to that of 2.4.6 is used to prove the above Theorems.

LEMMA 2.4.11 *Under the respective conditions of above Theorems 2.4.7 to 2.4.10, the estimate*

(2.4.10) $\qquad \int_0^\pi F(\varphi)\, P_n^{(\alpha+1,\ \beta)}(\cos\varphi)\, d\varphi \ = \ o(n^{-\alpha-1}), \quad -1 < \alpha \leq -\frac{1}{2},$

holds as $n \to \infty$.

Proof.　　Proof of the lemma is similar to the proof of lemma 2.4.6. As we proceed at first by breaking the interval of integration (2.4.10) into five parts

$$\int_0^{c/n} \ + \ \int_{c/n}^\delta \ + \ \int_\delta^{\pi-\delta'} \ + \ \int_{\pi-\delta'}^{\pi-c/n} \ + \int_{\pi-c/n}^\pi \ = \ \Sigma_{i=1}^5 T_i \ \text{(say)}.$$

Now, T_1 and T_2 are estimated using the orders (1.1.3) and the condition (2.4.9). Again in T_3 inequality (1.3.5) is used and Remman-Lebesgue lemma is applied to get

(2.4.11) $\qquad\qquad\qquad\qquad T_3 \ = \ o\!\left(n^{-\alpha-1}\right).$

Now, After changing the variable of integration and using the orders (1.3.1a) for small δ' and $n \to \infty$, we have

(2.4.12) $\qquad\qquad T_4 = \int_{\pi-\delta'}^{\pi-c/n} F(\varphi) P_n^{(\alpha+1,\ \beta)}(\cos\varphi)\, d\varphi$

$$= \ O(n^{-\frac{1}{2}}) \int_{c/n}^{\delta'} \left| \, f(-\cos\varphi) - A \, \right| \ \varphi^{2\beta+1} \ \varphi^{-\beta-1/2} \, d\varphi$$

$$= \ O(n^{-\frac{1}{2}}) \int_{c/n}^{\delta'} \ \varphi^{\beta+1/2} \left| \, f(-\cos\varphi) \right| \, d\varphi \ + \ o(n^{-\frac{1}{2}})$$

$$= \ o\!\left(n^{-\alpha-1}\right), \qquad \left(\, as \ \alpha \ \leq -\frac{1}{2} \ \text{so that} \ \alpha + \frac{1}{2} \ \leq \ 0\,\right),-$$

by the use of the 'antipole' conditions. Similar arguments are applied for T_5. This completes the proof of Lemma 2.4.11.

Proof of Theorems 2.4.7 to 2.4.10　　In the light of the lemma 2.4.11, the required proofs run as that of Theorems 2.4.1 to 2.4.3. Thus the proofs are complete

CHAPTER III

SUMMABILITY CASES OF F-J SERIES

3.1 Cesàro summability

Apart from the summability theorems at end points given in Chapter II, there is further advancement in the subject dealing the problems of convergence of summability sequences at the end points $x = \pm 1$. Cesàro summability of negative orders of an infinite series includes the ordinary convergence of that series. Regarding Jacobi series, Szegö[1] includes all results done upto 1967. Important works done are Rau([1, 1929],[2, 1936]), Obrechkoff([1, 1936]) and Kogbetliantz([1, 1919], [2, 1931]) etc. These are explained in Theorems 2.1.1, 2.1.2, 2.1.3 and 2.1.4 in Chapter II of this Monograph. Later works done in this direction are helpful to prove approximation problems by Jacobi polynomials and wavelets. Following Cesàro summability Theorems by Yadav[12] are being given to tame the approximation of functions by Jacobi polynomials.

THEOREM 3.1.1   *If $f(x)$ is continuous at $x = +1$, then the F-J series is (C, 1)-summable to $f(1)$ at $x = 1$, provided that $-1 < \alpha < 1/2, -1 < \beta \leq -1/2$ or $(\beta > -1/2$ but $\alpha + \beta \leq 0)$.*

THEOREM 3.1.2   *If $f(x)$ is continuous at $x = +1$, then the F-J series is (C, 1)-summable to $f(1)$ at $x = 1$, for $-1 < \alpha < 1/2$, $\beta > -1$; provided that the 'antipole' condition:*

(3.1.1) $\int_0^t \varphi^\beta \mid f(-\cos\varphi) \mid d\varphi = o(t^{\alpha-1}),$ as $t \to +0,$

is satisfied.

THEOREM 3.1.3   *If $f(x)$ is continuous at $x = +1$, then the F-J series is (C, 1)-summable to $f(1)$ at $x = 1$, for $-1 < \alpha < 1/2, \beta > -1$; provided that in case $\beta > -\frac{1}{2}$ and $\alpha + \beta > 0$;*

the additional 'antipole' condition:

(3.1.2) $\int_0^t \varphi^{\beta+1/2} \mid f(-\cos\varphi) \mid d\varphi = o(t^{\alpha-1/2}),$ as $t \to +0,$

is satisfied. (For $-1 < \beta \leq -1/2$ or for $\beta > -1/2$ but $\alpha + \beta \leq 0$, no antipole condition is necessary as proved in Theorem 3.1.1).

Following Lemma is used for the proof of above three Theorems.

LEMMA 3.1.4   *Under the conditions of each Theorem, we have*

(3.1.3) $\int_0^\pi F(\varphi)\, P_n^{(\alpha+2,\,\beta)}(\cos\varphi)\, d\varphi = \begin{cases} o(n^{-\alpha}), -1 < \alpha < \frac{1}{2} \\ o(n^{-\frac{1}{2}}\log n), \quad \alpha = \frac{1}{2} \\ o(n^{-\frac{1}{2}}), \quad \alpha > \frac{1}{2}, \end{cases}$

where

(3.1.4) $\qquad F(\varphi) = \{f(\cos\varphi) - A\}(\sin\varphi)^{2\alpha+1}(\cos\varphi)^{2\beta+1}$

REMARK 3.1.5   Continuity of f at $x = 1$ in the above three theorems can be replaced by a more general condition

(3.1.5) $\qquad \int_0^t |\, f(\cos\varphi) - A\,|\, (\sin\varphi)^{2\alpha+1}(\cos\varphi)^{2\beta+1}\, d\varphi = o(t^{2\alpha+2})$ as $t \to 0+$,

or in other format

(3.1.6) $\qquad \int_0^t |\, f(\cos\varphi) - A\,|\, \varphi^{2\alpha+1} d\varphi = o(t^{2\alpha+2})$ as $t \to 0+$,

or

$$\int_0^t |F(\varphi)|\, d\varphi = o(t^{2\alpha+2}) \quad \text{as } t \to 0+.$$

If the condition (3.1.6) is satisfied then (C, 1)-mean of F-J series will converge to A for which the end point order of (3.1.6) holds at $x = +1$. This condition is weaker than the continuity as shown in the remark 2.3.9.

Proof of the Lemma 3.1.4

Interval of integration in (3.1.3) is broken into five parts like that of (2.4.5) and named them T_1 to T_5. Using (1.3.1a), by the 'pole' condition (3.1.6),

$$T_1 = O(n^{\alpha+2}) \int_0^{c/n} |F(\varphi)|\, d\varphi = o(n^{-\alpha})$$

Again, let $F_1(\varphi) \equiv \int_0^\varphi |F(t)|\, dt$, then $\quad \forall\, \varepsilon > 0, \exists\, \delta > 0 : |F_1(\varphi)| < \varepsilon\, \varphi^{2\alpha+2}$ whenever $\frac{c}{n} < \varphi < \delta$ by the pole condition (3.1.6). Thus

$$T_2 = O(n^{-\frac{1}{2}}) \int_{c/n}^\delta \varphi^{-\alpha - 5/2} |F(\varphi)|\, d\varphi$$

$$= O(n^{-\frac{1}{2}}) \left[\left| \varphi^{-\alpha - \frac{5}{2}} |F_1(\varphi)| \right|_{c/n}^\delta + O(1) \int_{c/n}^\delta \varphi^{-\alpha - 7/2}\, \varepsilon\, \varphi^{2\alpha+2}\, d\varphi \right]$$

$$= \begin{cases} o(n^{-\alpha}), -1 < \alpha < \frac{1}{2} \\ o(n^{-\frac{1}{2}} \log n), \quad \alpha = \frac{1}{2} \\ o(n^{-\frac{1}{2}}), \qquad \alpha > \frac{1}{2}, \end{cases}$$

for ε arbitrary and $n \to \infty$. These two orders are the same for identical pole conditions in above three theorems. Again in T_3 inequality (1.3.5) is used and Remman-Lebesgue lemma is applied to get $o(n^{-1/2})$ which is distributed as orders required in (3.1.3). For T_4 and T_5 we examine each case separately for every Theorem.

Case (i) $-1 < \beta \le -1/2$.

$$T_5 = O(n^{-\frac{1}{2}}) \int_{\pi - c/n}^{\pi} |F(\varphi)| \, d\varphi$$

$$= o(n^{-1/2}), \quad as \ F \in L_1(0, \pi) \ and \ n \to \infty.$$

Also

$$T_4 = O\left(n^{-\frac{1}{2}}\right) \int_{\pi - \delta'}^{\pi - c/n} |F(\varphi)| \, d\varphi$$

$$= o\left(n^{-\frac{1}{2}}\right)$$

by the same argument as above where we use the relevant order of $P_n^{(\alpha + 2, \, \beta)}(\cos \varphi)$ valid in $[\pi/2, \pi]$ extracted by (1.3.2).

Case (ii) $\boldsymbol{\beta} > -1/2$, but $\boldsymbol{\alpha} + \boldsymbol{\beta} \leq \mathbf{0}$.

Since, we are considering the interval $[\pi/2, \pi]$. So

$$T_4 = O(n^\beta) \int_{\pi - \delta'}^{\pi - c/n} |F(\varphi)| \, d\varphi$$

$$= o(n^{-\alpha}) \, n^{\alpha + \beta} = o(n^{-\alpha}),$$

(for $\alpha + \beta \leq 0, \delta'$ arbitrarily small and $n \to \infty$). Also

$$T_5 = O(n^\beta) \int_{\pi - c/n}^{\pi} |F(\varphi)| \, d\varphi = o(n^{-\alpha}) \, n^{\alpha + \beta} = o(n^{-\alpha}).$$

Case (iii) $\boldsymbol{\beta} > -1$ with the antipole condition (3.1.1).

$$T_4 = \int_{c/n}^{\delta'} |F(\pi - \varphi)| \, P_n^{(\beta, \, \alpha+1)}(\cos \varphi) d\varphi$$

$$= O(n^{-1/2}) \int_{c/n}^{\delta'} \varphi^{-\beta - 1/2} \, |f(-\cos \varphi) - A|\varphi^{2\beta+1} \, d\varphi,$$

(by (1.3.1a))

$$= o(n^{-1/2}) + o(n^{-\alpha}), \quad \text{(by integrating by parts, using the antipole}$$

condition (3.1.1))

$$= \begin{cases} o(n^{-\alpha}), & (for - 1 < \alpha < \frac{1}{2}) \\ o(n^{-1/2}), & (\ for \ \alpha > \frac{1}{2}). \end{cases}$$

$$T_5 = \int_0^{c/n} |F(\pi - \varphi)| \, P_n^{(\beta, \, \alpha+1)}(\cos \varphi) d\varphi$$

$$= O(n^\beta) \int_0^{c/n} |f(-\cos \varphi) - A|\varphi^{2\beta+1} \, d\varphi,$$

$$= O(n^{-1}) \int_0^{c/n} \varphi^\beta \, |f(- \cos \varphi) - A| d\varphi = o(n^{-\alpha}).$$

(by (3.1.1))

Case (iv) $\beta > -1/2$, with $\alpha + \beta > 0$ and the antipole condition (3.1.2).

$$T_4 = \int_{c/n}^{\delta'} |F(\pi - \varphi)| \, P_n^{(\beta, \, \alpha+1)}(\cos \varphi) d\varphi$$

$$= O(n^{-1/2}) \int_{c/n}^{\delta'} \varphi^{\beta+1/2} \, |f(- \cos \varphi) - A| \, d\varphi,$$

By the 'antipole' condition (3.1.2), $\forall \varepsilon > 0, \exists \delta' > 0$:

$$\int_0^{\delta'} \varphi^{\beta+1/2} \, | \, f(\cos \varphi) \, | d\varphi < \varepsilon (\delta')^{\alpha-1/2},$$

whenever $\frac{c}{n} < \varphi < \delta'$. Thus

$$T_4 = O\left(n^{-\frac{1}{2}}\right) [\varepsilon (\delta')^{\alpha-\frac{1}{2}} - \varepsilon (c/n)^{\alpha-1/2}]$$

$$= o(n^{-1/2}) + o(n^{-\alpha}), \text{ (for small } \delta' \text{ and } n \to \infty).$$

$$= \begin{cases} o(n^{-\alpha}), \ (for - 1 < \alpha < \frac{1}{2}) \\ o(n^{-1/2}), \qquad (for \ \alpha \geq \frac{1}{2}). \end{cases}$$

$$T_5 = O(n^{-1/2}) \int_0^{c/n} \varphi^{\beta+1/2} \, |f(- \cos \varphi) - A| \, d\varphi,$$
$$= o(n^{-\alpha}),$$

(for $\beta > -1/2$ and the antipole condition).

We note that we don't use the sum $\alpha + \beta > 0$ to get these orders. So the Lemma for Theorem 3.1.3 holds as well. This proves the Lemma 3.1.4.

Proof of Theorems 3.1.1, 3.1.2 *and* 3.1.3. If we denote by $S_n(f, 1, X)$, the nth partial sum of F-J series (2.1.8) at x = 1 then for any constant A, due to orthogonality of Jacobi polynomials

(3.1.7) $\quad S_n(f, 1, X) - A = \int_0^\pi F(\varphi) \, h_n^{(\alpha+1, \, \beta)} P_n^{(\alpha+1, \, \beta)}(\cos \varphi) \, P_n^{(\alpha+1, \, \beta)}(1) \, A_n d\varphi$

where $A_n = a_1 n^{-1} + a_2 n^{-2} + \ldots + O(n^\lambda)$, ($\lambda > 0$ and a_i being constants). Now summing by parts ,

(3.1.8) $\qquad \frac{1}{n+1} \Sigma_{k=0}^n \, [S_k(f, 1, X) - A] =$

$$= \int_0^\pi \frac{1}{n+1} \sum_{k=0}^{n-1} \Delta(A_k) \frac{\Gamma(k+\alpha+\beta+3)}{2^{\alpha+\beta+2}\Gamma(\alpha+2)\Gamma(k+\beta+1)} F(\varphi) P_k^{(\alpha+2,\beta)}(\cos\varphi)\, d\varphi$$

$$+ \int_0^\pi \frac{2^{-\alpha-\beta-2}}{n+1}(A_n)\frac{\Gamma(n+\alpha+\beta+3)}{\Gamma(\alpha+2)\Gamma(n+\beta+1)} F(\varphi) P_n^{(\alpha+2,\beta)}(\cos\varphi)\, d\varphi$$

$$\leq \frac{1}{n+1}\sum_{k=0}^{n-1} |\Delta(A_k)| \frac{\Gamma(k+\alpha+\beta+3)}{2^{\alpha+\beta+2}\Gamma(\alpha+2)\Gamma(k+\beta+1)} \left| \int_0^\pi F(\varphi) P_k^{(\alpha+2,\beta)}(\cos\varphi)\, d\varphi \right|$$

$$+ \frac{2^{-\alpha-\beta-2}}{n+1}|A_n| \frac{\Gamma(n+\alpha+\beta+3)}{\Gamma(\alpha+2)\Gamma(n+\beta+1)} \left| \int_0^\pi F(\varphi) P_n^{(\alpha+2,\beta)}(\cos\varphi)\, d\varphi \right|$$

$$= O(n^\alpha)\, |o(n^{-\alpha})| = o(1),$$

as $n \to \infty$, by lemma 3.1.4 . This holds for all three theorems. This proves the Theorems.

A negative order Cesàro summability which includes the convergence of the series, has been proved by Pandey[1] given below.

THEOREM 3.1.6 *If the function*

$$(3.1.9) \qquad f(\omega) \equiv \{f(\omega) - A\}$$

belongs to $lip^*(\alpha + \tfrac{1}{2} - k)$, $\alpha - \tfrac{1}{2} < k < \alpha + \tfrac{1}{2}$, $-\tfrac{1}{2} < \alpha < +\tfrac{1}{2}$, $\beta \geq \alpha$. *Then the Fourier- Jacobi series is (C, k)-summable at x = 1.*

Following Theorem at x = 1 of F-J series is due to Gupta and Chaudhary[1].

THEOREM 3.1.7 *If*

$$(3.1.10) \qquad \int_0^t |F(\varphi)|\, d\varphi = o(n^{\alpha+3/2}), \text{ as } t \to 0+$$

Then the F-J series is summable by Hormonic mean at x = +1 to the sum A for - 1 < α < - ½, β > - ½ provided that the 'antipole' condition

$$(3.1.11) \qquad \int_{-1}^b (1+x)^{\frac{\beta}{2} - 3/4} |f(x)|\, dx < \infty$$

is satisfied for b fixed where

$$(3.1.12) \qquad F(\varphi) = \{f(\cos\varphi) - A\}(\sin\varphi)^{2\alpha+1}(\cos\varphi)^{2\beta+1}$$

It is to be noted that Theorems 3.1.6 and 3.1.7 do not provide any information about the case $\alpha \geq \beta \geq - \tfrac{1}{2}$. So we are not able to find out any approximation to an f by Jacobi polynomials. The end point Summability of F-J series at x = 1 for $\alpha \geq \beta \geq - \tfrac{1}{2}$ leads a process of approximation to a function on [- 1, +1]. Following Theorem due to Yadav [8] provides clue on such approximation.

THEOREM 3.1.8 *If*

(3.1.13) $\qquad \int_0^t |F(\varphi)| \, d\varphi = o(n^{2\alpha+2})$, as $t \to 0+$

Then the F-J series is (R, log n, 1)-summable to A at $x = +1$ for $-1 < \alpha < +\frac{1}{2}, \beta > -1$, provided that the antipole condition

(3.1.14) $\qquad \int_0^h \varphi^\beta \left| f(-\cos\varphi) \right| d\varphi = o(h^{\alpha-1})$, as $h \to 0+$,

holds. (R, log n, 1) mean of F-J series is defined by

(3. 1.15) $\qquad L(f, x, X) \equiv \sum_{k=0}^n \dfrac{\{S_k(x) - A\}}{\{log \ (n+1)\}(k+1)}$, $\qquad (n > 1)$

for $S_k(x)$ being partial sum of F-J series at $x \in [-1, +1]$.

Note: The condition (3.1.13) is weaker than the continuity of f at $x = +1$ (see the Remark 2.3.9). This Theorem 3.1.8 holds for $\alpha \geq \beta \geq -\frac{1}{2}$, so it is interesting to investigate an Approximation of a signal f by Jacobi polynomials and wavelets in the space X through *Logarithmic means* as we are going to present our works in Chapter IV and V. A conjecture for the said range of α, β may be

(3.1.16) $\qquad \left\| L(f,x,X) - f(x) \right\|_X \to 0,$

under the circumstances of the Theorem 3.1.8.

3.2 Absolute summabilities of F-J series

Scope of this Monograph is approximation of functions/signals which are p-power Lebesgue integrable with given weight. Absolute summability of related infinite series seems to be irrelevant. But the absolute summability of an infinite series implies the ordinary summability of the same order. Again the ordinary summability of F-J series at $x = +1$ leads a process of approximation on $[-1, +1]$ for $\alpha \geq \beta \geq -\frac{1}{2}$. It necessitate to at least mention the proved Theorems of absolute summability of F-J series at $x = +1$. Following Theorems are proved in Yadav[1] and [4]. We define a sequence of real numbers $\{p_n\}$ to be non-negative and non-increasing so that $\{p_n - p_{n-1}\}$ is non-increasing. Let

(3.2.1) $\qquad P_n = p_0 + p_1 + p_2 + \cdots + p_n$, $P_{-k} = p_{-k} = 0$ for $k = 1, 2, 3, \ldots$

and $P_n \neq 0$, $(n = 0, 1, 2, \ldots)$. Again, let $\{m_n\}$ be a positive monotone decreasing sequence such that

(3.2.2) $\qquad \sum_{n=1}^\infty \dfrac{m_n \ log \, n}{P_n} < \infty$

Then the Theorems in Yadav[1] and [4] are

THEOREM 3.2.1 *The series*

(3.2.3) $\qquad \sum_{n=0}^\infty m_n a_n P_n^{(\alpha, \beta)}(\cos\vartheta)$

is $|N, p_n|$*-summable at* $\theta = 0$ *provided that*

(3.2.4) $\qquad\qquad \varphi(\omega) \in lip(\alpha + \frac{1}{2}), \quad -\frac{1}{2} < \alpha < +\frac{1}{2}, \beta \geq \alpha;$

where

(3.2.5) $\qquad\qquad \varphi(\omega) = \{f(\cos \omega) - A\}(\sin \omega/2)^{\alpha + 1/2}(\cos \omega/2)^{\beta + 1/2}$

A is an absolute constant may be depending upon f.

THEOREM 3.2.2 *If* $\{\lambda_n\}$ *is non-negative non-increasing sequence such that* $\sum_{n=1}^{\infty} \lambda_n / n$ *is convergent. Then the series*

(3.2.6) $\qquad\qquad \sum_{n=0}^{\infty} \lambda_n \ a_n \ P_n^{(\alpha, \ \beta)}(\cos \vartheta)$

is absolutely summable (C, δ) *or summable* $|C, \delta|$ *at* $\theta = 0$, $(1 < \delta < 2)$ *for* $-\frac{1}{2} < \alpha < +\frac{1}{2}, \beta \geq \alpha;$ *provided that*

(3.2.7) $\qquad\qquad F(\omega) \equiv \{f(\cos \omega) - A\} \in lip(2 - \delta).$

REMARK 3.2.3 Theorem 3.2.1 implies the ordinary (N, p_n)-summability of the series (3.2.3). This means the sequence

(3.2.8a) $\qquad\qquad T_n(f, \cos \vartheta, X) \equiv \sum_{k=0}^{n} \frac{p_{n-k}}{p_n} \sum_{i=0}^{k} m_i \ a_i \ P_i^{(\alpha, \ \beta)}(\cos \vartheta)$

will converge to A at $x = \cos \vartheta = +1$. This leads an approximation of $f \in X$ i.e. a thorough investigation for the result

(3.2.8b) $\qquad\qquad \|T_n(f, \cos \vartheta, X) - f(\cos \vartheta)\|_X \to 0,$

is interestingly possible. Again from Theorem 3.2.2 convergence of Cesàro mean of order δ

(3.2.8c) $\qquad\qquad C_n^\delta(f, \cos \vartheta, X) \equiv \sum_{k=0}^{n} \frac{A_{n-k}^{\delta-1}}{A_n^\delta} \sum_{i=0}^{k} \lambda_i \ a_i \ P_i^{(\alpha, \ \beta)}(\cos \vartheta)$

is implied so that the investigation of an approximation of $f \in X$

(3.2.8d) $\qquad\qquad \|C_n^\delta(f, \cos \vartheta, X) - f(\cos \vartheta)\|_X \to 0,$

is interesting. Here to make $\alpha \geq \beta \geq -\frac{1}{2}$ one has to investigate some antipole conditions to avoid $\beta \geq \alpha.$

We write

(3.2.10) $\qquad\qquad F(\omega) \equiv \{f(\cos \omega) - A\}$

where A is an absolute constant and

(3.2.11) $\qquad \varphi(\omega) = \{F(\omega)\}(1 - \cos \omega)^{\alpha/2\,+\,1/4}(1 + \cos \omega)^{\beta/2\,+\,1/4}$

By BV[0, π], we mean bounded variation in the interval [o, π]. Then the following Theorems are proved in Yadav[6].

THEOREM 3.2.4   *The F-J series is absolutely convergent at* $x = +1$ *for* $-1 < (\alpha, \beta) < -\tfrac{1}{2}$, *provided that* $\varphi(\omega) \in BV[0, \pi]$.

THEOREM 3.2.5   *The F-J series is absolutely summable (C, k), (k > 0) for* $\alpha = \beta = -\tfrac{1}{2}$ *at* x $= +1$, *provided that* $F(\omega) \in BV[0, \pi]$ *and*

(3.2.12) $\qquad \int_0^t |F(\omega)| \, d\omega = O(t^2), \text{ as } t \to 0+.$

THEOREM 3.2.6   *The F-J series is absolutely convergent at* $x = +1$, *for* $-1 < (\alpha, \beta) < -\tfrac{1}{2}$, *provided that*

(3.2.13) $\qquad F(\omega) \in \text{lip } \delta/2 \ \text{ in } [0, \pi] \text{ for } \delta = \max(-\alpha - \tfrac{1}{2}, -\beta - 1/2)$

and

(3.2.14) $\qquad F(\omega) \in BV[0, \pi].$

if the antipole condition

(3.2.15) $\qquad \int_0^t \omega^\beta |f(-\cos \omega)| \, d\omega = O(t^{1/2}), \text{ as } t \to 0+$

is satisfied. If $\beta > \alpha$, *no antipole condition is necessary.*

THEOREM 3.2.7   *The F-J series is absolutely summable (C, 1), for* $\alpha = \beta = -\tfrac{1}{2}$ *at* $x = +1$, *provided that*

(3.2.16) $\qquad F(\omega) \in \text{lip } \delta, \ (\delta > 0).$

Note: Remark 3.2.3 is also applicable in these Theorems. That is the results can be used for approximation by polynomials and wavelets by managing the condition $\alpha \geq \beta \geq -\tfrac{1}{2}$. It is to be noted that these results holds good at other end point $x = -1$ only exchange in pole and antipole conditions and α, β are apparent. Theorems 3.2.4 to 3.2.7 are proved by a single technique given below.

Proof of Theorems 3.2.4 to 3.2.7.

The nth term of the Fourier-Jacobi series (2.1.2) at $\theta = 0$ is given by

(3.2.17) $\quad u_n = k_n \int_0^\pi F(\omega) \, (\sin \omega/2)^{2\alpha\,+\,1} (\cos \omega/2)^{2\beta\,+\,1} P_n^{(\alpha,\,\beta)}(\cos \omega) \, d\omega$

where

$$k_n = 2^{\alpha\,+\,\beta\,+\,1}(2n + \alpha + \beta + 1)\binom{n+\alpha}{n}\frac{\Gamma(n + 1)\Gamma(n+\alpha+\beta+1)}{\Gamma(n+\alpha+1)\Gamma(n+\beta+1)} \cong n^{\alpha\,+1}.$$

We break the interval of integration in (3.2.17) into intervals (0, c/n), (c/n, π – c/n) and (π – c/n, π) and denote them by I_1, I_2, and I_3 respectively. By the use of orders (1.3.2a), (1.3.2b), (1.1.5) and the antipole conditions, we have

(3.2.18) $\qquad\qquad I_1$ and $I_3 = O(n^{\alpha-1/2})$.

In I_2 we substitute (1.3.6) and apply the well known result for bounded variation of $F(\omega)$ and $\varphi(\omega)$ to get

$$I_2 = O(n^{\alpha+1/2}) \left| \int_{c/n}^{\pi-c/n} \varphi(\omega) \cos(N\omega + \gamma) \, d\omega \right| + O(n^{\alpha-1/2}) \int_{c/n}^{\pi-c/n} \frac{|\varphi(\omega)|}{\omega} \omega$$

(3.2.19) $\qquad\qquad\qquad\qquad = O(n^{\alpha-\frac{1}{2}} \log n)$.

This proves $\sum |u_n| < \infty$. Similar calculation is applied for $S_n(1)$, the partial sum of Jacobi series at $\cos\theta = +1$ and one has

(3.2.20) $\qquad\qquad\qquad\qquad \sum \frac{S_n(1)}{n^\delta} < \infty$

which proves $|C, \delta|$ summability of F-J series. This completes the required proofs. Here in all Theorems $\alpha \geq \beta \geq -\frac{1}{2}$ is not covered but the line of approach is open. Following Theorem due to Yadav [5] can be applied for approximation by wavelets. The strategy is that done in Chapter IV and V of this Monograph.

THEOREM 3.2.8 *The F-J series is summable* $|C, 1|$ *at* $x = +1$ *provided that*

(3.2.21) $\qquad F(\omega) \in lip \, \delta, \ (\delta > \alpha + 1/2, \ -1/2 < \alpha < 1/2; \ \beta > -1).$

and the antipole condition: that the integral

(3.2.22) $\qquad\qquad \int_0^t \omega^{\beta-1/2} |f(-\cos\omega)| \, d\omega, \ (t > 0),$

exists. For $\beta \geq \alpha$, *no antipole condition is required.*

Proof of Theorem 3.2.8 At $x = +1$, we have to apply the result $\sum \frac{S_n(1)}{n} < \infty$ to prove the $|C, 1|$-summability. For this I_1 and I_3 are disposed of as in Theorem 3.2.7. I_2 is summarized in the following way:

Real and imaginary parts of the integral

(3.2.23) $\qquad \int_{c/n}^{\pi-c/n} (\sin\omega/2)^{\alpha-1/2} (\cos\omega/2)^{\beta+1/2} F(\omega) \cos(N\omega + \gamma) \, d\omega$

makes it equal to

(3.2.24)

$$\frac{1}{2}\left\{\int_{\frac{c}{n}}^{\pi-\frac{c}{n}} - \int_{\frac{c}{n}-\mu_n}^{\pi-\frac{c}{n}-\mu_n}\right\} \times$$

$$\cdot \left\{\left(\sin\frac{\omega}{2}\right)^{\alpha-\frac{1}{2}} (\cos \omega/2)^{\beta+\frac{1}{2}} F(\omega) \, e^{i\,\omega\,(2n+\alpha+\beta+2)/2}\right\} d\omega$$

where $\mu_n = \dfrac{\pi}{n + \frac{\alpha+\beta+2}{2}}$. Its real and imaginary parts are calculated by breaking the intervals into four parts

$$\tfrac{1}{2}(L_1 + L_2 + L_3 + L_4), \text{ (say)};$$

where

$$L_1 \leq \left|\int_{c/n}^{\pi-c/n}\right| = O(n^{-\alpha-\frac{1}{2}-\delta}),$$

$$L_2 \leq \left|\int_{\pi-\frac{c}{n}-\mu_n}^{\pi-c/n}\right| = O(n^{-1}),$$

$$L_3 \leq \left|\int_{c/n}^{\pi-\frac{c}{n}-\mu_n}\right| = O(n^{-\delta}),$$

and

$$L_4 \leq \left|\int_{c/n}^{\pi-\frac{c}{n}-\mu_n}\right| = O(n^{-1} \log n).$$

Without loss of generality, we have set A $=$ $f(1) = 0$ to get

(3.2.25) $F(\omega) = O(1 - \cos \omega) = O(\omega^2).$

Since $F(\omega) \in \text{lip } \delta$, so that

(3.2.26) $F(\omega + \mu_n) - F(\omega) = O(\mu_n)^\delta.$

Collecting the orders and substituting them, we get

$$\sum \frac{S_n(1)}{n} < \infty,$$

which proves the Theorem. For details one may consult Yadav[5].

CHAPTER IV

APPROXIMATION BY JACOBI POLYNOMIALS

4.1 Saturation classes of approximation processes

(1) Introduction A series termed as F-J expansion associated with function $f \in X$ called signal is given by (2.1.8) and is written in the form

(4.1.1) $$f(\cos \vartheta) \sim \sum_{k=0}^{\infty} \hat{f}(k) \, \omega_k^{(\alpha, \, \beta)} R_k^{(\alpha, \, \beta)}(\cos \vartheta)$$

where coefficients are defined by (2.1.6) and (2.1.7). If we denote a translate of $f(x)$ (usually denoted by $f(x + t)$) by $T_\varphi f(\cos \theta)$ then it is proved (see Askey and Waingar[1]), that

(4.1.2) $$T_\varphi f(\cos \vartheta) \sim \sum_{k=0}^{\infty} \hat{f}(k) \, \omega_k^{(\alpha, \, \beta)} R_k^{(\alpha, \, \beta)}(\cos \vartheta) \, R_k^{(\alpha, \, \beta)}(\cos \varphi)$$

such that

$$\|T_\varphi f\|_X \leq \|f\|_X$$

and

$$s - \lim_{\varphi \to 0} T_\varphi \, f \, = \, f$$

$T_\varphi f$ is called generalized translate of f on $[-1, +1]$. It is known that the operator T_φ is positive operator for $\alpha \geq \beta \geq -\frac{1}{2}$ with operator norm 1 (see Gasper[1]). Also

(4.1.3) $$R_k^{(\alpha, \, \beta)}(\cos \vartheta) \, R_k^{(\alpha, \, \beta)}(\cos \varphi) = \int_0^\pi K(\cos \vartheta, \cos \varphi, \cos \psi) \, R_k^{(\alpha, \, \beta)}(\cos \psi)$$

$$\cdot \, \rho^{(\alpha, \, \beta)}(\psi) \, d\psi,$$

where

(4.1.4) $$\rho^{(\alpha, \, \beta)}(\psi) \equiv w(\cos \psi) \equiv (1 - \cos \psi)^\alpha \, (1 + \cos \psi)^\beta \sin \psi$$

$$= 2^{\alpha + \beta + 1}(\sin \psi/2)^{2\alpha + 1} \cos \psi/2)^{2\beta + 1}$$

The function $K(\cos \vartheta, \cos \varphi, \cos \psi) \geq 0$, is symmetric function in $\cos \vartheta, \cos \varphi$ and $\cos \psi$ for $\alpha \geq \beta \geq -\frac{1}{2}$ and

(4.1.5) $$\int_0^\pi K(\cos \vartheta, \cos \varphi, \cos \psi) \, \rho^{(\alpha, \, \beta)}(\psi) \, d\psi \, = 1.$$

Moreover,

(4.1.6) $$T_\psi f(\cos \theta) = \int_0^\pi f(\cos \varphi) \, K(\cos \vartheta, \cos \varphi, \cos \psi) \, \rho^{(\alpha, \, \beta)}(\varphi) \, d\varphi$$

Let $S_n(f, \cos \vartheta, X) \equiv S_n(f, \cos \vartheta)$ be the nth partial sum of the series (4.1.1). Then

$$S_n(f, \cos \vartheta) = \sum_{k=0}^{n} \hat{f}(k)\, \omega_k^{(\alpha, \beta)}\, R_k^{(\alpha, \beta)}(\cos \vartheta)$$

$$= \sum_{k=0}^{n} \int_0^{\pi} f(\cos \varphi)\, \omega_k^{(\alpha, \beta)}\, \{ R_k^{(\alpha, \beta)}(\cos \varphi)\, R_k^{(\alpha, \beta)}(\cos \vartheta)\}\, \rho^{(\alpha, \beta)}(\varphi) d\varphi$$

$$= \sum_{k=0}^{n} \int_0^{\pi} f(\cos \varphi)\, \omega_k^{(\alpha, \beta)}\, \{ \int_0^{\pi} K(\cos \vartheta, \cos \varphi, \cos \psi)$$
$$\cdot R_k^{(\alpha, \beta)}(\cos \psi)\, \rho^{(\alpha, \beta)}(\psi)\, d\psi \}\, \rho^{(\alpha, \beta)}(\varphi)\, d\varphi$$

(by (4.1.3)),

$$= \int_0^{\pi} f(\cos \varphi)\, \{ \int_0^{\pi} K(\cos \vartheta, \cos \varphi, \cos \psi)$$
$$\sum_{k=0}^{n} \omega_k^{(\alpha, \beta)}\, R_k^{(\alpha, \beta)}(\cos \psi)\, \rho^{(\alpha, \beta)}(\psi)\, d\psi \}\, \rho^{(\alpha, \beta)}(\varphi)\, d\varphi$$

$$= \int_0^{\pi} L_n\, R_n^{(\alpha + 1, \beta)}(\cos \psi)\, \rho^{(\alpha, \beta)}(\psi)\, d\psi$$
$$\times \int_0^{\pi} f(\cos \varphi)\, K(\cos \vartheta, \cos \psi, \cos \varphi)\, \rho^{(\alpha, \beta)}(\varphi)\, d\varphi$$

(by symmetricity of K and summation (1.2.6)),

$$(4.1.7) \qquad = L_n \int_0^{\pi} T_\psi f(\cos \theta)\, R_n^{(\alpha+1, \beta)}(\cos \psi)\, \rho^{(\alpha, \beta)}(\psi)\, d\psi,$$

(by (4.1.6)), where

$$L_n = \frac{\Gamma(n + \alpha + \beta + 2)}{\Gamma(\alpha + 1)\, \Gamma(n + \beta + 1)}\, P_n^{(\alpha+1, \beta)}(1)$$

$$= \frac{(\alpha + 1)\, \omega_n^{(\alpha+1, \beta)}}{2n + \alpha + \beta + 2} = n^{2\alpha + 2}\, \{ 1 + O(1/n) \}$$

$$= n^{2\alpha + 2}\, L(n) \qquad (\text{say}),$$

It is to be recoded that $L(n)$ is slowly varying function of n in the sense that for some real numbers ε, $\delta > 0$, there are natural numbers n_1 and n_2 such that $n^{\varepsilon} L(n)$ is increasing from $n > n_1$ and $n^{-\delta} L(n)$ is decreasing from $n > n_2$. By the orthogonality of Jacobi polynomials

$$(4.1.8) \quad S_n(f, \cos \vartheta) - f(\cos \theta))$$

$$= L_n \int_0^{\pi} [T_\psi f(\cos \theta) - f(\cos \theta)]\, R_n^{(\alpha+1, \beta)}(\cos \psi)\, \rho^{(\alpha, \beta)}(\psi)\, d\psi$$

If we consider a lower triangular matrix $\Lambda \equiv (\lambda_{n,k})$, ($k = 0, 1, 2, \ldots, n$) with $\lambda_{n,0} = 1$ for all n. Then Λ − transform of the series (4.1.1) is defined by

$$(4.1.9) \qquad \sigma_n^\Lambda (f,\ \cos\vartheta) = \sum_{k=0}^n \lambda_{n,\ n-k}\ \hat{f}(k)\omega_k^{(\alpha,\ \beta)}\ R_k^{(\alpha,\ \beta)}(\cos\vartheta)$$

$$= \sum_{k=0}^n (\Delta\lambda_{n,\ n-k})\ S_k(f,\ \cos\vartheta)$$

$$= \sum_{k=0}^n L_k\ (\Delta\lambda_{n,\ n-k}) \int_0^\pi T_\psi f(\cos\theta)\ R_k^{(\alpha+1,\ \beta)}(\cos\psi)\ \rho^{(\alpha,\ \beta)}(\psi)\ d\psi$$

where $\Delta\lambda_{n,\ n-k} = \lambda_{n,\ n-(k+1)} - \lambda_{n,\ n-k}$ and thus $\sum_{k=0}^n \Delta\lambda_{n,\ n-k} = \lambda_{n,\ 0} = 1$,

while $\Delta\lambda_{n,\ k} = \lambda_{n,\ k} = 0$ for $n < k$. Thus

$$(4.1.10) \qquad \sigma_n^\Lambda (f,\ \cos\vartheta)\ - f(\cos\theta)$$

$$= \sum_{k=0}^n L_k\ (\Delta\lambda_{n,\ n-k}) \int_0^\pi [T_\psi f(\cos\theta) - f(\cos\theta)]\ R_k^{(\alpha+1,\ \beta)}(\cos\psi)\ \rho^{(\alpha,\ \beta)}(\psi)\ d\psi$$

$$= \int_0^\pi [T_\psi f(\cos\theta) - f(\cos\theta)] \left\{ \sum_{k=0}^n L_k\ (\Delta\lambda_{n,\ n-k})\ R_k^{(\alpha+1,\ \beta)}(\cos\psi) \right\} \rho^{(\alpha,\ \beta)}(\psi)\ d\psi$$

$$(4.1.11) \qquad = \int_0^\pi [T_\psi f(\cos\theta) - f(\cos\theta)]\ \{ K_n^\Lambda(\psi) \} \rho^{(\alpha,\ \beta)}(\psi)\ d\psi \qquad \text{(say)},$$

The kernel $K_n^\Lambda(\psi)$ is called '*kernel of σ^Λ process of approximation*' or in short '*the kernel*".

(2) Modulus of continuity

The function $\omega(\varphi,\ f, X)$ defined by

$$(4.1.12) \qquad \omega(\varphi,\ f, X) \equiv \omega(\varphi) = \sup_{0\ \le\ \psi\ \le\ \varphi} \| T_\psi f(.) - f(.) \|$$

(where $T_\psi f$ is generalized translate of f on $[-1, +1]$ defined by (4.1.2)), is called the modulus of continuity of $f \in X$. Also, we say $f \in Lip\ (\gamma,\ X)$, ($0 < \gamma \le 2$) if there exists a positive number c such that

$$(4.1.13) \qquad \omega(\varphi,\ f, X) \le c\ \varphi^\gamma,$$

The class of functions $Lip\ (\gamma,\ X)$ is a Banach space if endowed with the norm

$$(4.1.14) \qquad \|f\|_{Lip\ (\gamma, X)} = \|f\|_X + \sup_{0\ \le\ \psi\ \le\ \varphi} (n^{-1}\ \omega(n^{-1}, f, X))$$

For all these details see Bavinck[1].

(3) **Approximation processes**

We study the norm

(4.1.15) $$\| \sigma_n^A (f, \ cos\,\vartheta) \ - \ f(cos\,\theta) \|$$

in the space X with matrices $((\Delta\lambda_{n,\ n-k}))$ and $((\Delta\lambda_{n,\ k}))$. (see Yadav[9],[10] [16]). The numerical value of (4.1.15) is called order of σ_n^A –process of approximation. If the value of (4.1.15) tends to zero as n tends to infinity, we say that f can be replaced by $\sigma_n^A f$ or f is represented by $\sigma_n^A f$ to any degree of accuracy on $[-1, +1]$. The two matrices $(\lambda_{n,\ k})$ and $(\Delta\lambda_{n,\ k})$ are primary and secondary matrices and are related by the equation

(4.1.16) $$\lambda_{n,\ k} = \begin{cases} 1 - \sum_{v=0}^{k-1}\Delta\lambda_{n,\ k}, & k \le n \\ 0 & , \ k > n \end{cases}$$

where

(4.1.17) $$\Delta\lambda_{n,\ k} = \lambda_{n,\ k+1} - \lambda_{n,\ k} \ .$$

This suggests how one can sort out if either of any is known. Some commonly known matrices are,

(4.1.18) (i) $$\Delta\lambda_{n,\ n-k} = \begin{cases} \frac{1}{n+1}, & k \le n \\ 0, & k > n \end{cases}$$

(used for (C, 1)) mean).

(ii) $$\Delta\lambda_{n,\ n-k} = \begin{cases} A_{n-k}^{\mu-1}/A_n^{\mu}, & k \le n \\ 0, & k > n \end{cases}$$

(used for (C, μ)) mean).

(iii) $$\Delta\lambda_{n,\ n-k} = \begin{cases} \frac{p_{n-k}}{P_n}, & k \le n \\ 0, & k > n \end{cases}$$

(used for (N, p_n) mean).

(iv) $$\Delta\lambda_{n,\ n-k} = \begin{cases} \frac{p_k}{P_n}, & k \le n \\ 0, & k > n \end{cases}$$

(used for ($\bar{N}, p_n$) mean).

(v) $\qquad \Delta\lambda_{n,\;n-k} = \begin{cases} \dfrac{1}{(k+1)\;\Sigma_{j=0}^{n}\frac{1}{j+1}} \;, & k \leq n \\[4mm] 0, & k > n \end{cases}$

(used for (R, log n, 1) mean).

(vi) $\qquad \Delta\lambda_{n,\;n-k} = \begin{cases} 1, & k = n \\ 0, & k \neq n \end{cases}$

(used for identity transformation).

Also the matrices

(vii) $\qquad \Delta\lambda_{n,\;k} = \begin{cases} \dfrac{1}{(k+1)^{\mu}\;\Sigma_{j=0}^{n}\frac{1}{(j+1)^{\mu}}} \;, & k \leq n \\[4mm] 0, & k > n \end{cases}$

(viii) $\Delta\lambda_{n,\;k} = \begin{cases} \dfrac{1}{n+1}\left\{ 2 - (n+1)/\left\{ (k+1)\dfrac{1}{(k+1)\;\Sigma_{j=0}^{n}\frac{1}{j+1}}\right\}\right\}', & k \leq n \\[4mm] 0, & k > n \end{cases}$

(ix) $\qquad \Delta\lambda_{n,\;k} = \begin{cases} (\dfrac{1}{n-k+1})/\left\{\Sigma_{j=0}^{n}\frac{1}{j+1}\right\}, & k \leq n \\[4mm] 0, & k > n \end{cases}$

and many more matrices fall under the generality of our works which can be used for approximation.

(4) Concept of approximation

Calculating the partial sum of F-J series in the way as have been done in (4.1.7) one gets at $x = \cos\theta = 1$, (i.e. $\theta = 0$),

(4, 1, 19) $\qquad \sigma_n^{\Lambda}\,(f,\;1) - A = \Sigma_{k=0}^{n}\,(\Delta\lambda_{n,\;n-k})\,\{\;S_k(f,\;1) - A\;\}$

$\qquad\qquad = \left\{ \int_0^{\pi}[f(\cos\varphi) - A]\;\Sigma_{k=0}^{n}\;L_k\,(\Delta\lambda_{n,\;n-k})\;R_k^{(\alpha+1,\;\beta)}(\cos\varphi)\;\right\}\rho^{(\alpha,\;\beta)}(\varphi)\,d\varphi$

$\qquad\qquad = \int_0^{\pi}[f(\cos\varphi) - A]\;\{K_n^{\Lambda}\,(\varphi)\;\}\,\rho^{(\alpha,\;\beta)}(\varphi)\,d\varphi$

where A is an absolute constant, may depend upon f . But the two important relations described in (4.1.2) namely

$$\left\|T_{\varphi}f\right\|_X \;\leq\; \|f\|_X$$

and

$$s - lim_{\varphi \to 0} \, T_\varphi \, f \, = \, f$$

together imply that

(4. 1. 20) $$\left\| T_\varphi f \, - \, f \right\|_X \cong \, \left\| f \, - \, A \right\|_X$$

(see Yadav[18]). This in turn explains that boundedness of either involves the other and

(4.1.21) $$\left\| \, \sigma_n^A \, (f, \, 1) - \, A \, \right\|_X \, \to \, 0 \, \text{ as } \, n \, \to \, \infty,$$

implies

(4.1.22) $$\left\| \sigma_n^A \, (f, \, cos \, \vartheta) \, - \, f(cos \, \theta) \, \right\|_X \, \text{ as } \, n \, \to \, \infty$$

for $cos \, \vartheta \in [-1, +1]$. This statement holds only when $\alpha \geq \beta \geq -\frac{1}{2}$. The transformation $\sigma_n^A f$ is called an approximation to or strong approximation of $f \in X$. Following theorem may be used as a folklore of Approximation in mathematical analysis.

THEOREM   4.1.1   *If a sequence $\left\{ \sigma_n^A \, (f, \, x) \right\}$ of linearly transformed F-J series converges to A at $x = +1$, then the linear combination $\sigma_n^A \, (f, \, x)$ is an approximate form of $f(x)$ when $\alpha \geq \beta \geq -\frac{1}{2}$ at any point $x \in [-1, +1]$ and represents the function f to any degree of accuracy.*

(5) Saturation orders

Let in the space X, the difference between $\sigma_n^A f$ and f be given by

(4.1.23) $$\left\| \sigma_n^A \, (f, \, cos \, \vartheta) \, - \, f(cos \, \theta) \, \right\|_X \, \leq \, O_n$$

for a nontrivial elements f of X and tends to zero as $n \to \infty$. Again if there exists at least one nontrivial element for which

(4.1.24) $$\left\| \sigma_n^A \, (f, \, cos \, \vartheta) \, - \, f(cos \, \theta) \, \right\|_X \, > \, O_n.$$

Then the order O_n is called *Saturation Order* of the *Approximation Process* σ_n^A. The class of functions for which the order (4.1.23) holds and O_n tends to zero is called *Saturation class* or *Favard's class* of the process σ_n^A.

(6) Convolution in X

A type of product called convolution denoted by * is defined and its algebraic properties are deduced in Askey and Waingar[1]. If $f_1 , \, f_2 \in X$ then

(4.1.25) $$(f_1 * f_2)(cos \, \vartheta) \, = \, \int_0^\pi T_\varphi f_1(cos \, \theta) \, f_2(cos \, \theta)\} \, \rho^{(\alpha, \, \beta)}(\varphi) \, d\varphi$$

where $T_\varphi f$ is generalized translate of f on $[-1, +1]$ given by (4.1.2). Its algebraic properties when f_1, f_2 and $f_3 \in L^1(w)$ and $g \in X$ are

(i) $$f_1 * f_2 = f_2 * f_1$$

(ii) $$(f_1 * f_2) * f_3 = f_1 * (f_2 * f_3)$$

(iii) $$\| f_1 * g \|_X \leq \|f_1\|_{L_1} \|g\|_X$$

and

(iv) $$(\widehat{f_1 * f_2})(n) = (\widehat{f_1})(n) \, (\widehat{f_2})(n)$$

Thus the transform $\sigma_n^\Lambda f$ given in (4.1.10) can be written as

(4.1.26)
$$\sigma_n^\Lambda (f, \cos\vartheta) = (f * K_n^\Lambda)(\cos\vartheta)$$
$$= (K_n^\Lambda * f)(\cos\vartheta)$$

and

$$\|\sigma_n^\Lambda (f, \cos\vartheta) \|_X \leq \|f\|_X \|K_n^\Lambda\|_{L_1}$$

After the preparation of the above terminology, we state and prove the following Theorems on saturation given in Yadav[9] [10] where the restriction (3.2) in Yadav[9] is avoidable.

THEOREM 4.1.2 *Let* $((\Delta\lambda_{n,\ n-k}))$ *be a lower triangular matrix with* $\sum_{k=0}^{n} \Delta\lambda_{n,\ k} = \lambda_{n,\ 0} = 1$. *Let the partial sequence* $\{(k+1)^{\alpha+\frac{1}{2}} \Delta\lambda_{n,\ k}\}$, $(k \leq n)$ *be non-negative non-decreasing. Then the saturation order of* $\Lambda - process$ *of approximation is given by*

(4.1.27) $$\left\| f(.) - \sigma_n^\Lambda (f, (.)) \right\|_X \equiv \left\| f(.) - (f * K_n^\Lambda)(.) \right\|_X$$

$$\leq A \left(\sum_{k=0}^{n} \frac{\omega(\frac{1}{k+1})}{(k+1)^{\alpha+3/2}} \sum_{v=0}^{k} \Delta\lambda_{n,\ n-v} \ (n-v)^{\alpha+\frac{1}{2}} L(n-v) + n^{-2\beta-2} \right)$$

for $\alpha \geq \beta \geq -\frac{1}{2}$. *The saturation class or the Favard's class is collection of those f in X for which, the right hand side of (4.1.27) tends to zero as* $n \to \infty$.

The last term in (4.1.27) can be absorbed in the first summation, if the modulus of continuity is restricted by

(4.1.28) $$\omega(\tfrac{1}{k+1}) \geq \begin{cases} A\,(k+1)^{\alpha-2\beta-\frac{3}{2}}(n-k+1)^{-\alpha-1/2} \\ or \\ B(k+1)^{\alpha-2\beta-\frac{1}{2}}(n-k+1)^{-\alpha-3/2} \end{cases}$$

for A, B absolute constants, n large enough and $k = 0, 1, 2, ..., n$; with $\alpha \geq \beta \geq -\frac{1}{2}$. The following Theorem is refined in one term.

THEOREM 4.1.3 *Let* $((\Delta\lambda_{n,\ n-k}))$ *be as in Theorem 4.1.2 with (4.1.28) satisfied. Then*

(4.1.29) $$\left\| f(.) - \sigma_n^\Lambda (f, (.)) \right\|_X \equiv \left\| f(.) - (f * K_n^\Lambda)(.) \right\|_X$$

$$\le A\left(\sum_{k=0}^{n}\frac{\omega(\frac{1}{k+1})}{(k+1)^{\alpha+3/2}}\sum_{v=0}^{k}\Delta\lambda_{n,\;n-v}\;(n-v)^{\alpha+\frac{1}{2}}L(n-v)\right)$$

and the saturation classes are as those given in Theorem 4.1.2.

If we denote by $S_n^1(f,\;cos\,\vartheta)$, the (C, 1) mean of the F-J series. Then the following corollary of Theorem 4.1.3 holds by writing $\Delta\lambda_{n,\;n-k}=1/(n+1)$ for $k\;\le\;n$ and zero otherwise in Theorem 4.1.3.

COROLLARY   4.1.4   For $\alpha\ge\beta\ge-\tfrac{1}{2}$,

(4.1.30) $$\left\|f(.)-\;s_n^1(f,\;(.))\right\|_X\;\le\;A\,n^{\alpha-1/2}\sum_{k=0}^{n}\frac{\omega(\frac{1}{k+1})}{(k+1)^{\alpha+1/2}}$$

whenever the restrictions of Theorem 4.1.3 hold.

Also, taking $\Delta\lambda_{n,\;n-k}=1$ for $k\;=\;n$ and zero otherwise in Theorem 4.1.3, we get the following result for partial sum of F-J series in X.

COROLLARY   4.1.5   For $\alpha\ge\beta\ge-\tfrac{1}{2}$, if (4.1.28) holds then, we have

(4.1.31) $$\left\|S_n(f,\;(.))\right\|_X\;\le\;\begin{cases}A\,n^{\alpha+1/2},\;\alpha>-1/2\\A\,log\,n,\;\alpha=-1/2\end{cases}$$

where S_n denote the nth partial sum of F-J series at any point in $[-1,+1]$.

The triangular matrix $((\Delta\lambda_{n,\;n-k}))$ defines the following transformation in view of

$$\sum_{k=0}^{\infty}\Delta\lambda_{n,\;n-k}\equiv\sum_{k=0}^{\infty}\Delta\lambda_{n,\;k}\;=\;1;$$

that

(4.1.32)

$$\sigma_n^\Lambda(f,\;cos\,\vartheta)-f(cos\,\theta)\equiv A\sum_{k=0}^{n}(k+1)\{S_k^1(f,\;cos\,\vartheta)-f(cos\,\vartheta)\}\Delta^2\lambda_{n,\;k}.$$

Then applying the corollary 4.1.4, we have

COROLLARY   4.1.6   Let $((\Delta\lambda_{n,\;n-k}))$ be as in Theorem 4.1.2 with (4.1.28) satisfied. Then

(4.1.33) $\left\|f(.)-\;\sigma_n^\Lambda(f,\;(.))\right\|_X$

$$\le A\left(\sum_{k=0}^{n}|\Delta^2\lambda_{n,\;k}|\,(k+1)^{\alpha+\frac{1}{2}}\sum_{v=0}^{k}\frac{\omega(\frac{1}{v+1})}{(v+1)^{\alpha+1/2}}\right)$$

and the saturation classes are as those given in Theorem 4.1.2.

Proof of Theorem 4.1.2

The interval of integration in (4.1.10) is broken into $(0, \frac{\pi}{n+1})$, $(\frac{\pi}{n+1}, \pi - \frac{\pi}{n+1})$ and

$(\pi - \frac{\pi}{n+1}, \pi)$ and are denoted by P, Q and R respectively. Now

$$P \leq A \int_0^{\frac{\pi}{n+1}} \omega(\psi, f) \, |K_n^A(\psi)| \, \rho^{(\alpha, \beta)}(\psi) \, d\psi$$

$$\leq A \; \omega\left(\frac{1}{n}\right) n^{-2\alpha - 2} \sum_{v=0}^n \Delta\lambda_{n, \, n-v} \, L(n - v)$$

$$(4.1.34) \qquad \leq A \left(\sum_{k=0}^n \frac{\omega(\frac{1}{k+1})}{(k+1)^{\alpha+3/2}} \sum_{v=0}^k \Delta\lambda_{n, \, n-v} \, (n - v + 1)^{\alpha+\frac{1}{2}} \, L(n - v) \right)$$

$$Q \leq A \int_{\frac{\pi}{n+1}}^{\pi - \frac{\pi}{n+1}} \omega(\psi, f) \left| \left\{ \sum_{k=0}^{n_0} + \sum_{n_0}^n \right\} L_k \, \Delta\lambda_{n, \, n-k} \, R_k^{(\alpha+1, \, \beta)}(\cos \psi) \right|$$
$$\cdot \rho^{(\alpha, \beta)}(\psi) \, d\psi$$

where n_0 is fixed number such that the Hilb's type formula is valid in $(\frac{\pi}{n+1}, \pi - \frac{\pi}{n+1})$ for n > n_0 (see Bingham [4]). The first sum is bounded for fixed n_0 and the idea that $\Delta\lambda_{n, \, k} \to 0$ as n $\to \infty$ for fixed k , is applied. In the second sum a result of Mc Fadden [1] which reads

$$(4.1.35) \qquad \left| \sum_a^b \Delta\lambda_{n, \, n-k} \, e^{i \, (n-k)t} \right| \leq (2\pi + 1) \sum_{k=0}^{[\frac{1}{\psi}]} \Delta\lambda_{n, \, n-k} \, ,$$

$(0 \leq a \leq b \leq \infty)$ is used to get

$$Q \leq A \left(\sum_{k=0}^n \int_{\frac{\pi}{k+2}}^{\frac{\pi}{k+1}} \frac{\omega(\psi)}{(\psi)^{\frac{1}{2} - \alpha}} \sum_{v=0}^{[\frac{1}{\psi}]} \Delta\lambda_{n, \, n-v} \, (n - v + 1)^{\alpha+\frac{1}{2}} \, L(n - v) \right) d\psi$$

for all $n = 0, 1, 2, \ldots$. The order (4.1.27) follows on integration and getting out the maximum values for $[\frac{k+2}{\pi}] \leq$ k. Also for

$$R \leq A \int_{\pi - \frac{\pi}{n+1}}^{\pi} \omega(\psi, f) \, |K_n^A(\psi)| \, \rho^{(\alpha, \beta)}(\psi) \, d\psi,$$

we see that

$$(4.1.36) \qquad K_n^A(\psi) \equiv F(\cos \psi)$$

$$= \Delta\lambda_{n,\ 0}\, L_0\ R_0^{(\alpha+1,\ \beta)}(\cos\psi) + \sum_{k=1}^{n} b_k\ \omega_k^{(\alpha+1,\ \beta)}\, R_k^{(\alpha+1,\ \beta)}(\cos\psi)$$

with

(4.1.37)
$$b_k = \frac{(\alpha+1)\Delta\lambda_{n,\ n-k}}{2 + \frac{\alpha+\beta+2}{k}}\ ,$$

is bounded and monotonic. Bavinck[1] has proved that $F(\cos\psi)$ is continuous and uniformly bounded in $0 < \psi \leq \pi$. Thus

(4.1.38)
$$R \leq An^{-2\beta-2}.$$

This completes the proof of Theorem 4.1.2.

Proof of Theorem 4.1.3   The last term in (4.1.27) is absorbed if the restriction (4.1.28) on the modulus of continuity is imposed. Thus the proof of Theorem 4.1.3 is completed on the lines of the proof of Theorem 4.1.2. This completes the proof of Theorem 4.1.3.

Proof of corollaries of Theorems 4.1.2 and 4.1.3 are particular instances of the situations in the theorems. Another face of the idea of approximation that involves non-negative non-increasing trend of the partial sequence $\{k^{\alpha+1/2}\Delta\lambda_{n,\ k}\}$, ($k \leq n$) has also been considered by Yadav[10] and following results are proved.

THEOREM   4.1.7   *Let* $((\Delta\lambda_{n,\ k}))$ *be a lower triangular matrix such that* $\sum_{k=0}^{n}\Delta\lambda_{n,\ k} = \lambda_{n,\ 0}$ $= 1$ *and the partial sequence* $\{k^{\alpha+1/2}\Delta\lambda_{n,\ k}\}$, *($k \leq n$) be non-negative non-increasing. Then for* $\alpha \geq \beta \geq -\ \frac{1}{2}$,

(4.1.39) $\left\| f(.) - \sigma_n^\Lambda\left(f,\ (.)\right) \right\|_X \equiv \left\| f - \left(f\ *\ K_n^\Lambda\right)\right\|_X$

$$\leq A\left(\sum_{k=0}^{n} \frac{\omega\left(\frac{1}{k+1}\right)}{(k+1)^{\alpha+3/2}}\ \sum_{v=0}^{k}\Delta\lambda_{n,\ v}\ (v+1)^{\alpha+\frac{1}{2}}\, L(v+1) + n^{-2\beta-2}\right)$$

whence the transformation $\sigma_n^\Lambda\left(f,\ (.)\right)$ *defined by (4.1.9) contains* $((\Delta\lambda_{n,\ k}))$ *in place of* $((\Delta\lambda_{n,\ n-k}))$. *For* $\omega\left(\frac{1}{n}\right) \geq$ A $n^{-2\beta-2}$, *the sum in (4.1.39) absorbs the last term. Right side inequality gives the saturation order of the process* σ_n^Λ. *Favard's class is collection of those* $f \in X$ *for which the order in (4.1.39) tends to zero.*

There are many important particular cases of Theorem 4.1.7. Following case $\alpha = \beta = -\ \frac{1}{2}$ is interesting.

COROLLARY   4.1.8   *Let* $S_n^1(f,\ \cos\vartheta, X) \equiv S_n^1(f, \cos\vartheta)$ *denote the (C, 1) mean of F-J series. Then for* $\alpha = \beta = -\ \frac{1}{2}$,

(4.1.40) $\left\| f(.) - s_n^1\left(f,\ (.)\right) \right\|_X \leq An^{-1}\sum_{k=0}^{n}\ \omega\left(\frac{1}{k+1}\right)$

The orders obtained are Saturation orders. Results similar to (4.1.33) are

(4.1.41)
$$\left\| f(.) \ - \ \sigma_n^A \left(f, \ (.) \right) \right\|_X \leq A \left(\Sigma_{k=0}^n |\Delta^2 \lambda_{n, \ k}| \ \Sigma_{v=0}^k \ \omega(\tfrac{1}{v+1}) \right)$$

for $\alpha = \beta = - \frac{1}{2}$.

Proof.   Order (4.1.40) is a particular case of (4.1.39) when substituted $\Delta \lambda_{n, \ k} = 1/(n+1)$ for k $\leq$ n and zero otherwise with $\alpha = \beta = - \frac{1}{2}$. Again to obtain (4.1.41) one substitutes (4.1.40) in the transform (4.1.32) where case $\alpha = \beta = - \frac{1}{2}$ is allowed. This completes the proof.

Note: The Process (4.1.39) allows the (R, log n, 1) mean matrix and the matrix

(4.1.42)
$$\Delta \lambda_{n, \ k} = \begin{cases} \dfrac{1}{(k+1)^\mu \ \Sigma_{j=0}^n \frac{1}{(j+1)^\mu}}, & k \leq n \\ \\ 0, & k > n \end{cases}$$

for $- \frac{1}{2} \leq \alpha \leq + \frac{1}{2}$ and $\alpha \geq \beta \geq - \frac{1}{2}$ (i.e. $+ \frac{1}{2} \geq \alpha \geq \beta \geq - \frac{1}{2}$) when μ is adjusted with non-increasingness. These processes permit different types of signals to be replaced by Jacobi polynomials and then by wavelets. In these regards, the results obtained are unique.

Proof of Theorem 4.1.7   To complete the proof of this Theorem, one has to proceed on the lines of the proof of Theorem 4.1.2 and get the orders for P and R as those are there. In the assessment of Q we use the result similar to that of Mc Fadden[1] that

(4.1.43)
$$\left| \Sigma_a^b \Delta \lambda_{n, \ k} \ (k+1)^{\alpha+1/2} \ e^{ikt} \ dt \right| \ \leq \ A \ \Sigma_{k=0}^{[\frac{1}{t}]} \Delta \lambda_{n, \ k} \ (k+1)^{\alpha+1/2} \ ,$$

$(0 \leq a \leq b \leq \infty)$, $0 < t < \pi, 0 \leq k \leq n$, proved by Yadav[10]. This completes the necessary proof.

4.2   Some special instances of approximation

Previous section elaborates the saturation orders in terms of modulus of continuity which gives different classes of functions for different orders of modulus. Certain conditional approximations to an $f \in X$ in terms of Jacobi polynomials are proved and even more are to be searched in coming days. At the same hand it is perhaps certain but to be proved that every convergence order obtained by regular summability method is a saturation order. Some important approximation results already proved are being given in this section. Following Theorems due to Yadav[16] holds. We rewrite the conditions for the sake of clarity of the idea tolerating the fallacy of repetition.

THEOREM   4.2.1   *If* $f \in X$ ($\frac{1}{2} > \alpha \geq \beta \geq - \frac{1}{2}$, *but in case* $\beta > - \frac{1}{2}$ *we must have* $\alpha + \beta$ ≤ 0). *Then f can be represented by a linear combination of Jacobi polynomials, say* Cesàro mean of order one σ_n^1 (f, $\cos \vartheta$) *on* [-1, +1] *to any degree of accuracy* i.e. *for* $\cos \vartheta \in [-1, +1]$

(4.2.1)
$$\| f(\cos \vartheta) \ - \ \sigma_n^1 \ (f, \ \cos \vartheta) \|_X \ \to \ 0 \ \text{as } n \to \infty,$$

provided that the 'pole' condition

(4.2.2)
$$\int_0^t |F(\varphi)| \ d\varphi \ = \ o(t^{2\alpha+2}), \ \text{as } \ t \to 0+$$

48

where

(4.2.3) $$F(\varphi) = \{f(\cos\varphi) - A\}(\sin\varphi/2)^{2\,\alpha+1}(\cos\varphi/2)^{2\beta\,+\,1}$$

is satisfied for A an absolute constant depending upon f . $\sigma_n^1 f$ is Fezér mean or (C, 1) mean of F-J series.

In a further advancement Yadav[19] again has proved the following Theorem which covers the case $\beta > -\frac{1}{2}$ but $\alpha + \beta > 0$ by adding an additional "antipole" condition.

THEOREM   4.2.2   *Let $f \in X$ and satisfy the 'pole' condition (4.2.2) and let for $\beta > -\frac{1}{2}$ but $\alpha + \beta > 0$, the antipole condition*

(4.2.4) $$h^{\frac{1}{2}\,-\,\alpha}\int_0^h t^{\beta+1/2}\,|f(-\cos t)|\,dt \;=\; o(1),\,as\;h\to\;0+$$

hold. Then the relation (4.2.1) is true for $\frac{1}{2} > \alpha \geq \beta \geq -\frac{1}{2}$.

It is interesting to note from the following Theorem proved in Yadav[14] that the two cases ($\beta > -\frac{1}{2}$ but $\alpha + \beta \leq 0$) and ($\beta > -\frac{1}{2}$ but $\alpha + \beta > 0$) are covered by a single '*antipole*' condition:

(4.2.5) $$\int_0^h \varphi^\beta\,|f(-\cos\varphi)|\,d\varphi \;=\; o(h^{\alpha-1}),\,as\;h\to\;0+$$

for $\beta > -1$. The said Theorem is

THEOREM   4.2.3   *The statement (4.2.1) holds for both the cases ($\beta > -\frac{1}{2}$ but $\alpha + \beta \leq 0$) and ($\beta > -\frac{1}{2}$ but $\alpha + \beta > 0$) provided that in addition to the pole condition (4.2.2), the 'antipole' condition (4.2.5) is satisfied for $\beta > -1$. For the first case ($\beta > -\frac{1}{2}$ but $\alpha + \beta \leq 0$) no antipole condition is necessary.*

Before we reach the latest achievement, we discus the proof of Theorems 4.2.1, 4.2.2 and 4.2.3. We keep in mind the fact of the Theorem 4.1.1 which states that convergence of every linearly transformed sequence of F-J series at x = 1   leads a process of approximation for the signals by Jacobi polynomials on [-1, + 1] when $\alpha \geq \beta \geq -\frac{1}{2}$. Thus for all the cases of above theorems following result holds.

LEMMA   4.2.4   *Under the circumstances of Theorems 4.2.1, 4.2.2 and 4.2.3, the (C, 1) mean of F-J series denoted by $\sigma_n^1(f,\cos\vartheta)$ converges to A at $\vartheta = 0$ i.e.*

(4.2.6) $$\sigma_n^1(f,\,1) \;\to\; A,\;as\;n\;\to\;\infty;$$

or

(4.2.7) $$lim_{n\to\infty}\frac{1}{n+1}\sum_{k=0}^n\{S_k(1) - A\} \to\; 0,$$

for $-1 < \alpha < \frac{1}{2}$, $\beta > -1$ and A being an absolute constant depending upon f . $S_k(1)$ is partial sum of F-J series at $\vartheta = 0$.

Following Lemma is in the heart of the proof of lemma 4.2.4

LEMMA 4.2.5 *Under the said conditions of all the Theorems 4.2.1, 4.2.2 and 4.2.3 , we have the estimate*

(4.2.8)

$$\int_0^\pi F(\varphi)\, R_n^{(\alpha+2,\ \beta)}(\cos\varphi)\, d\varphi \; = \; \begin{cases} o(n^{-2\alpha-2}), & -1 < \alpha < +\tfrac{1}{2} \\ o(n^{-\alpha-\frac{5}{2}}\log n), & \alpha = +\tfrac{1}{2} \\ o(n^{-\frac{5}{2}}), & \alpha > +\tfrac{1}{2}, \end{cases}$$

for every respective range of β in the theorems.

Proof. The interval of integration in (4.2.8) is broken into five parts,

$$\int_0^{c/n} \; + \; \int_{c/n}^{\delta} \; + \; \int_{\delta}^{\pi-\delta'} \; + \; \int_{\pi-\delta'}^{\pi-c/n} \; + \; \int_{\pi-c/n}^{\pi} \; = \; \Sigma_{i=1}^5 T_i \quad (\text{say}).$$

The estimate of each part has thoroughly been explained in the publication Yadav[14] . The proof is parallel to that of Lemma 2.4.6 if one replaces the Jacobi polynomials $P_k^{(\alpha+2,\ \beta)}(\cos\varphi)$ by the normalized Jacobi polynomials

$$R_k^{(\alpha+2,\ \beta)}(\cos\varphi) \equiv \frac{\left\{ P_k^{(\alpha+2,\ \beta)}(\cos\varphi) \right\}}{\left\{ P_k^{(\alpha+2,\ \beta)}(1) \right\}}$$

Then applying the pole condition and order (1.3.1) the orders for T_1 and T_2 are obtained with the help of c, δ, *and* δ' which are arbitrarily small positive reals. Using the formula (1.3.5) with Riemann- Lebesgue lemma, we get,

$$T_3 = o(n^{-\alpha-5/2}) \; = \; \begin{cases} o(n^{-2\alpha-2}), & -1 < \alpha < +\tfrac{1}{2} \\ o(n^{-\alpha-\frac{5}{2}}\log n), & \alpha = +\tfrac{1}{2} \\ o(n^{-\frac{5}{2}}), & \alpha > +\tfrac{1}{2}, \end{cases}$$

Again the estimates of T_4 and T_5 are obtained in both the cases $(\beta > -\tfrac{1}{2}$ but $\alpha + \beta \leq 0)$ and $(\beta > -\tfrac{1}{2}$ but $\alpha + \beta > 0)$ with the applications of the antipole conditions (4.2.4) and (4.2.5). Hence the proof of the Lemma is complete.

Proof of the Lemma 4.2.4 Let us write $S_n(1)$ to denote the nth partial sum of the Jacobi series given in (2.1.8) for normal form. Then, we have

(4.2.12) $\qquad S_n(1) - A \; = \; \int_0^\pi F(\varphi)\, \omega_n^{(\alpha+1,\ \beta)}\, R_n^{(\alpha+1,\ \beta)}(\cos\varphi)\, c_n\, d\varphi$

by the orthogonality of Jacobi polynomials and summation formula of Szegö([1] page 71) given in (1.2.6), where

$$c_n = \frac{2(\alpha + 1)}{(2n + \alpha + \beta + 2)}$$

Thus the difference between (C, 1) mean of F-J series and an absolute constant A is

$$\sigma_n^1(f, 1) - A \equiv \frac{1}{n+1} \sum_{k=0}^n \{S_k(1) - A\}$$

$$= \int_0^\pi [f(\cos\varphi) - A] \left[\frac{1}{n+1} \sum_{k=0}^n \omega_k^{(\alpha+1,\ \beta)} R_k^{(\alpha+1,\ \beta)}(\cos\varphi) \right]$$

$$\left\{ \frac{2(\alpha+1)}{(2k + \alpha + \beta + 2)} \right\} \rho^{(\alpha,\ \beta)}(\varphi)\, d\varphi$$

By Abel's transformation

$$|\sigma_n^1(f, 1) - A| \leq \left| \int_0^\pi F(\varphi) \left[\frac{1}{n+1} \sum_{k=0}^{n-1} \Delta c_k \sum_{i=0}^k \omega_i^{(\alpha+1,\ \beta)} R_i^{(\alpha+1,\ \beta)}(\cos\varphi) \right] d\varphi \right|$$

$$+ \left| \int_0^\pi F(\varphi) \left[\frac{c_n}{n+1} \sum_{i=0}^n \omega_i^{(\alpha+1,\ \beta)} R_i^{(\alpha+1,\ \beta)}(\cos\varphi) \right] d\varphi \right|$$

(4.2.13)
$$\leq \left| \int_0^\pi F(\varphi) \left[\frac{1}{n+1} \sum_{k=0}^{n-1} \Delta c_k \frac{2(\alpha+2)}{(2k+\alpha+\beta+3)} \omega_k^{(\alpha+2,\ \beta)} R_k^{(\alpha+2,\ \beta)}(\cos\varphi) \right] d\varphi \right|$$

$$+ \left| \int_0^\pi F(\varphi) \left[\frac{c_n}{n+1} \frac{(\alpha+2)}{(2n+\alpha+\beta+3)} \omega_n^{(\alpha+2,\ \beta)} R_n^{(\alpha+2,\ \beta)}(\cos\varphi) \right] d\varphi \right|$$

$$= P_1 + P_2 \quad \text{(say)},$$

But for $\varepsilon > 0$, $\exists\ m : n \geq m$ implies,

$$\left| \int_0^\pi F(\varphi) R_k^{(\alpha+2,\ \beta)}(\cos\varphi)\, d\varphi \right| < \varepsilon\, k^{-2\alpha-2}$$

by an application of (4.2.8). So, we have

$$P_1 \leq \left[\frac{1}{n+1} \sum_{k=m}^{n-1} \Delta c_k \frac{(\alpha+2)}{(2k+\alpha+\beta+3)} \omega_k^{(\alpha+2,\ \beta)} \right] \varepsilon\, k^{-2\alpha-2} + A_1\varepsilon$$

$$< A_2 \left[\frac{1}{n+1} \varepsilon \sum_{k=m}^{n-1} k^{-3} k^{2\alpha+5} k^{-2\alpha-2} \right] + A_1\varepsilon$$

$$< A_3\varepsilon \quad (A_i \text{ are absolute constants}).$$

In the same way $P_2 < A_4\varepsilon$. Consequently it follows that

(4.2.14) $\sigma_n^1(f,1) \rightarrow A$ as n $\rightarrow \infty$,

for $-1 < \alpha < +\frac{1}{2}$ but β as in the respective Theorems. This proves the Lemma.

Proof of Theorems 4.2.1, 4.2.2 *and* 4.2.3. In view of the relation (4.1.20) we have for $\alpha \geq \beta \geq -\frac{1}{2}$,

$$\| \sigma_n^1(f, \cos\vartheta) - f(\cos\vartheta) \|_X \leq \| \sigma_n^1(f, 1) - A \|_X$$

But the right hand side tends to zero by (4.2.14). This completes the proofs of the Theorems mentioned.

4.3   Best possible cases of approximation

We studied the saturation classes in section §4.1. There the formulas explain the approximation orders in terms of modulus of continuity in $L_p^{(\alpha,\beta)}$ spaces of Lebesgue integrable functions/signals with weight w. Again in section §4.2 our searches concerned the approximation of a function by (C, 1)-mean of F-J series under sufficient conditions. Necessary parts are untouched in these sections. Now, we survey the general order Cesàro summablity of F-J series and find that there is a final boundary of approximation beyond which the statement of the representation of function in terms of Jacobi polynomials, is not true. To make the idea clear, we quote the relations of Cesàro means from Zygmund[1] for ready reference. Let $u_0 + u_1 + u_2 + \dots$, be an infinite series with partial sums S_k (k = 0, 1, 2, ...). It is customary to define the Cesàro mean of order k by $\sigma_n^k = S_n^k / A_n^k$ where S_n^k is called Cesàro sum of the infinite series and A_n^k is binomial number given by

(4.3.1) (i) $(1-x)^{-k-1} = \sum_{n=0}^{\infty} A_n^k x^n$,

$$\sum_{n=0}^{\infty} S_n^k x^n = \frac{\sum_{n=0}^{\infty} S_n x^n}{(1-x)^k} = \frac{\sum_{n=0}^{\infty} u_n x^n}{(1-x)^{k+1}} ,$$

where k is not any positive or negative integer. From (i) it follows that

(ii) $A_n^k = \binom{n+k}{n} = \frac{(k+1)(k+2) \dots (k+n)}{\llcorner(n)} = \frac{\Gamma(n+k+1)}{\Gamma(k+1)\Gamma(n+1)}$

$$\cong \frac{n^k}{\Gamma(k+1)}, \quad (k \neq -1, -2, \dots)$$

(iii) $A_n^{k_1+k_2+1} = \sum_{v=0}^{n} A_v^{k_1} A_{n-v}^{k_2}$

(iv) $S_n^{k_1+k_2+1} = \sum_{v=0}^{n} A_{n-v}^{k_1} S_v^{k_2}$

(v) $S_n^k = \sum_{v=0}^{n} A_{n-v}^{k-1} S_v = \sum_{v=0}^{n} A_{n-v}^k u_v$

(vi) $A_n^k = \sum_{v=0}^{n} A_v^{k-1} = \sum_{v=0}^{n} A_{n-v}^{k-1}, \quad A_n^k - A_{n-1}^k = A_n^{k-1}$

(vii) $S_n^k = \sum_{v=0}^{n} A_{n-v}^{k-1} S_v = \sum_{v=0}^{n} A_{n-v}^k u_v, \quad S_n^k - S_{n-1}^k = S_n^{k-1},$

also

$$S_n^k \;=\; \sum_{v=0}^{n} S_v^{k-1} \;=\; \sum_{v=0}^{n} S_{n-v}^{k-1}$$

(viii) $\qquad\qquad \sum_{v=0}^{\infty} \mid A_v^k \mid \;\; < \;\; \infty, \qquad$ for $k < -1$

(ix) $\quad A_n^k$ is positive for $k > -1$, increasing for $k > 0$, decreasing for $\;0 > k > -1$.
For $k < -1$, A_n^k is of constant sign.

These relations are very much helpful to understand the calculations in analysis. We denote by $\sigma_n^k(f, \cos\vartheta)$ the kth Cesàro mean of order k of F-J series at $\cos\vartheta \in [-1, +1]$. So,

$$\sigma_n^k(f, \cos\vartheta) \;=\; \left(\frac{S_n^k(\cos\vartheta)}{A_n^k} \right) \;=\; \frac{1}{A_n^k} \sum_{v=0}^{n} A_{n-v}^{k-1}\, S_v(\cos\vartheta)$$

and for any absolute constant A, by relation (4.3.1) (vi)

$$\sigma_n^k(f, \cos\vartheta) \;-\; A \;=\; \frac{1}{A_n^k} \sum_{v=0}^{n} A_{n-v}^{k-1}\, \{ S_v(\cos\vartheta) \;-\; A \,\}$$

$$= \frac{1}{A_n^k} \sum_{v=0}^{n} A_{n-v}^{k-1} \left\{ \sum_{i=0}^{v} \hat{f}(i)\, \omega_i^{(\alpha,\,\beta)}\, R_i^{(\alpha,\,\beta)}(\cos\vartheta) - A \,\right\}$$

(see (2.1.8)). At $\vartheta = 0$ i.e. $x = \cos\vartheta = 1$, $\;R_i^{(\alpha,\,\beta)}(\cos\vartheta) = 1$. Thus the above relation at the pole $x = +1$ is

$$\sigma_n^k(f, 1) \;-\; A \;=$$

$$= \frac{1}{A_n^k} \sum_{v=0}^{n} A_{n-v}^{k-1} \left[\sum_{i=0}^{v} \int_0^{\pi} \{ f(\cos\varphi) - A \,\}\, \omega_i^{(\alpha,\,\beta)}\, R_i^{(\alpha,\,\beta)}(\cos\varphi)\, \rho^{(\alpha,\beta)}(\varphi) d\,\varphi \right]$$

(substituting for $\hat{f}(i)$, using the orthogonal property of Jacobi polynomials for A)

$$= \frac{1}{A_n^k} \sum_{v=0}^{n} A_{n-v}^{k-1} \left[\int_0^{\pi} F(\varphi) \sum_{i=0}^{v} \omega_i^{(\alpha,\,\beta)}\, R_i^{(\alpha,\,\beta)}(\cos\varphi)\, d\varphi \right]$$

(using the notation (4.2.3) and changing the order of summation and integration)

$$(4.3.2) \qquad = \frac{1}{A_n^k} \sum_{v=0}^{n} A_{n-v}^{k-1} \left[\int_0^{\pi} F(\varphi) \frac{2(\alpha+1)}{2v+\alpha+\beta+2}\, \omega_v^{(\alpha+1,\,\beta)}\, R_v^{(\alpha+1,\,\beta)}(\cos\varphi) d\varphi \right]$$

(by summation formula (1.2.6)). It is clear that (4.3.2) is difference of (C, k) mean of F-J series at $\vartheta = 0$ and the absolute constant A. Also, by (4.1.10), the difference

$$[\sigma_n^k \ (f, \ cos \ \vartheta) \ - \ f(cos \ \theta)] \ =$$

$$\left[\int_0^\pi [T_\varphi f \ (cos \ \vartheta) - f(cos \ \theta)] \ \frac{1}{A_n^k} \Sigma_{\nu=0}^n \ A_{n-\nu}^{k-1} \ \frac{2(\alpha+1)}{2\nu+\alpha+\beta+2} \ \omega_\nu^{(\alpha+1, \ \beta)} \ R_\nu^{(\alpha+1, \ \beta)}(cos \ \varphi)d\varphi \right]$$

by writing

$$\Delta\lambda_{n, \ n-\nu} = \begin{cases} A_{n-\nu}^{k-1}/A_n^k, & \nu \leq n \\ 0, & \nu > n \end{cases}$$

in the equation (2.9) of Yadav[9] i.e in (4.1.10). Again by (4.1.20), it is clear that

(4.3.4) $\|\sigma_n^k \ (f, \ cos \ \vartheta) \ - \ f(cos \ \theta) \|_X \ \leq A \ \| \sigma_n^k \ (f, \ 1) - \ A \|_X$

A and A_i (i = 1, 2, 3, ...) are absolute constants but not the same at each occurrence. We are now ready to state an important Theorem proved in Yadav[18].

THEOREM   4.3.1   *Let f $\in$ X and $\alpha \geq \beta \geq$ - ½, then f can be represented by a linear combination of Jacobi polynomials say, $\sigma_n^k \ (f, \ cos \ \vartheta)$ in [- 1, +1] to any degree of accuracy i.e.*

(4.3.5) $\|\sigma_n^k \ (f, \ cos \ \vartheta) \ - \ f(cos \ \theta) \|_X \to 0 \ \ as \ \ n \to \infty+,$

(1) *if the 'pole' condition*

(4.3.6) $\int_0^t |f(cos \ \varphi) - \ A| \ \varphi^{2\alpha+1} \ d\varphi \ = \ o(t^{2\alpha+2}), \ \ as \ t \to 0+$

is satisfied and $k \geq \alpha+\beta+1$.

(2)   *For $\beta > $ - ½ , $\alpha + ½ \ < \ k \ < \ \alpha + \beta + 1$, the approximation (4.3.5) holds provided in addition to the 'pole' condition (4.3.6), the 'antipole' condition*

(4.3.7) $\int_0^h |f(- \ cos \ \varphi)| \ \varphi^{\beta-\alpha+k} \ d\varphi \ = \ o(1), \ \ as \ h \to 0+$

is satisfied.

(3)   *For $k \leq \alpha + ½$, the statement   (4.3.5) is not true in the sense that there exists a non-trivial function f in X such that the norm in (4.3.5) tends to infinity.*

(4)   *For $\alpha + ½ < k < \alpha + \beta + 1$ but without the 'antipole' condition (4.3.7), the statement (4.3.5) is not true in the sense that there exists a function f in X such that the norm in (4. 3. 5) tends to infinity.*

(5)   *It is noticeable that for $k \geq \alpha + \beta + 1$, no antipole condition is necessary.*

Proof of the Theorem 4.3.1 is based upon the following lemmas.

LEMMA   4.3.2   *Let f $\in$ X for $\alpha, \beta \ > $ - 1 such that the 'pole' condition (4.3.6) is satisfied. Then*

$$(4.3.8) \quad \int_0^\pi F(\varphi)\, R_n^{(\alpha+k+1,\ \beta)}(\cos\varphi)\, d\varphi = \begin{cases} o(n^{-2\alpha-2}), & k > \alpha + \tfrac{1}{2} \\ o(n^{-\alpha-k-3/2}\log n), & k = \alpha + \tfrac{1}{2} \\ o(n^{-\alpha-k-3/2}), & k < \alpha + \tfrac{1}{2} \end{cases}$$

as $n \to \infty$; where

$$(4.3.9) \quad F(\varphi) = \{f(\cos\varphi) - A\}\, (\sin\varphi/2)^{2\alpha+1}(\cos\varphi/2)^{2\beta+1}$$

provided that for $\beta > -\tfrac{1}{2}$, $\alpha + \tfrac{1}{2} < k < \alpha + \beta + 1$, the 'antipole' condition (4.3.7) is satisfied. For $-1 < \beta \le -\tfrac{1}{2}$ or (for $\beta > -\tfrac{1}{2}$, but $k \ge \alpha + \beta + 1$) no antipole condition is necessary.

Proof. The Lemma 2.3.10 proved in section 2.3 is in the usual form of Jacobi polynomial $P_n^{(\alpha,\ \beta)}(\cos\varphi)$. If one replaces that in the normal form of Jacobi polynomials $R_n^{(\alpha,\ \beta)}(\cos\varphi)$, the result and the proof becomes the same for the Lemma 4.3.2. Thus the proof of the Lemma 4.3.2 follows on the lines of proof of the lemma 2.3.10. This completes the proof of the Lemma 4.3.2.

LEMMA 4.3.3 *Let $f \in X$ and $\alpha, \beta > -1$. Then under the conditions of the Lemma 4.3.2, we have*

$$(4.3.10) \quad \sigma_n^k(f,\ 1) \to A \quad \text{as } n \to \infty.$$

Proof. From (4.3.2), it follows that

$$(4.3.11) \quad \left| \sigma_n^k(f,\ 1) - A \right| =$$

$$\left| \frac{2^{-\alpha-\beta-1}}{\Gamma(\alpha+1)} \frac{1}{A_n^k} \sum_{v=0}^n G_v(n,k)\, \frac{2v+\alpha+\beta+k+2}{\Gamma(v+\beta+1)}\, \frac{\Gamma(v+\alpha+\beta+k+2)\,\Gamma(v+\alpha+k+2))}{\Gamma(v+1)\,\Gamma(\alpha+k+2)} \right|$$

$$\cdot \left| \int_0^\pi F(\varphi)\, R_v^{(\alpha+k+1,\ \beta)}(\cos\varphi)\, d\varphi \right|,$$

where

$$(4.3.12) \quad G_v(n,k) = \sum_{m=v}^n A_{n-m}^k\, A_{m-v}^{-k-2}\, (2m+\alpha+\beta+1)\frac{\Gamma(m+v+\alpha+\beta+1)}{\Gamma(m+v+\alpha+\beta+k+3)}$$

(see Szegö[1] page 258). Thus by lemma 4.3.2

$$\left| \sigma_n^k(f,\ 1) - A \right| \le O(n^{-k}) \sum_{v=0}^n |G_v(n,k)|\, O(v^{(2\alpha+2k+3)})\, o(v^{-2\alpha-2})$$

$$(4.3.13) \qquad = o(n^{-k}) \sum_{v=0}^n |G_v(n,k)|\, o(v^{2k+1})$$

(4.3.14) $\qquad = M_n^{(k)} \qquad$ for $k > \alpha + \frac{1}{2}$ and under the conditions of the Lemma.

But

(4.3.15) $\qquad M_n^{(k)} \rightarrow 0$ as $n \rightarrow \infty$ (see Szegö [1] page 263).

This completes the proof of the Lemma 4.3.3.

Proof of Theorem 4.3.1 Proof of the main result (4.3.5) has been given in four theorems of Yadav[18]. A survey of the same is being presented. Since

(4.3.16) $\qquad \left\| \sigma_n^k (f, \cos \vartheta) - f(\cos \theta) \right\|_X \leq A \left\| \sigma_n^k (f, 1) - A \right\|_X$

for some absolute constant A. It is correct for any linear transformation $\sigma_n^\Lambda f$ by a lower triangular matrix Λ. Again, among the present circumstances of the Theorem

$$\| \sigma_n^k (f, 1) - A \|_X$$

tends to zero as proved in (4.3.10). So, the assertion (4.3.5) follows. To show that it is the best possible assertion, we proceed as follows.

For $k \leq \alpha + \frac{1}{2}$, the assertion (4.3.5) is not correct. It is known that there exist continuous functions (see Szegö[1] eqn. (9.41.17)) such that for $k = \alpha + \frac{1}{2}$

(4.3.17) $\qquad \sigma_n^k (f, 1) > C \log n, \quad (C > 0)$

and by the regularity it follows that the statement (4.3.5) is not true for $k \leq \alpha + \frac{1}{2}$. This proves the unique range of k. Also, it is proved in Yadav[13] that for the function

$$(1 + x)^\mu, \quad (\mu > -\beta - 1)$$

the Jacobi series exists but the 'antipole' condition (4.3.7) is not satisfied for $-1 - \beta < \mu \leq \frac{1}{2} (\alpha - \beta - k - 1)$. It may readily be seen that $\mu \leq \frac{1}{2} (\alpha - \beta - k - 1) \Leftrightarrow 2\mu + \beta - \alpha + k \leq -1$ and for $f(x) = (1 + x)^\mu$ i.e. $f(-\cos \varphi) = (1 - \cos \varphi)^\mu$

$$\Rightarrow \quad \int_0^h |f(-\cos \varphi)| \, \varphi^{\beta - \alpha + k} \, d\varphi = A \int_0^h \varphi^{2\mu + \beta - \alpha + k} \, d\varphi$$

doesn't exist and the antipole condition (4.3.7) doesn't hold. The left inequality $\mu > -1 - \beta \Rightarrow \mu + \beta > -1$ explains the existence of F−J series. Also the principal term of the F-J series associated with the function $(1 + x)^\mu, \quad (\mu > -\beta - 1)$ is

(4.3.18) $\qquad = (-1)^n \, n^{\alpha - \beta - 2\mu - 1}$

(see Szegö[1] equation (9.42.12)) which is not (C, k) summable even for $\beta > -\frac{1}{2}$, $\alpha + \frac{1}{2} < k < \alpha + \beta + 1$. Thus the antipole condition is necessary. This shows the uniqueness of the antipole condition. We believe that with this discussion the best possible boundary of the approximation (4.3.5) in X is established. The proof is complete.

Following Theorem is the latest outcome of our study (Yadav[22]) on approximation.

THEOREM 4.3.4 *Let* $X_6^{\alpha,\beta} \subset X$, $(\alpha = \beta = -\tfrac{1}{2})$ *be a space of signals which satisfy the 'pole' condition*

(4.3.19) $\qquad \int_0^t \varphi^{\alpha-1/2} |f(\cos\varphi) - A| \, d\varphi = o(t^{\alpha+1/2})$ *as* $t \to 0+$.

Then

(4.3.20) $\qquad\qquad \|S_n(f, x) - f(x)\|_x \to 0, \qquad$ *as* $n \to \infty$.

Or $\qquad \forall \varepsilon > 0, \; \exists \, n_0 : n \geq n_0 \;\Rightarrow\; \|S_n(f, \cos\vartheta) - f(\cos\theta)\|_x < \varepsilon,$

where S_n *is partial sum of the F-J series. Or in other words: the signals of the subspace* $X_6^{\alpha,\beta}$ *can be represented by the partial sums of F-J series for* $\alpha = \beta = -\tfrac{1}{2}$ *to any degree of accuracy.*

Note: The pole condition (4.3.19) is not trivial. At the first hand it may be seen that this is satisfied by the signals of the class *Lip* δ, $(\delta \geq 1/4)$ and many more spaces are expected to be in the range of the Theorem 4.3.4 e.g. a class whose members have sink at the pole $x = +1$.

Proof of the Theorem 4.3.4 If we denote by $S_n(f,x)$ the partial sum of F-J series, then

(4.3.21) $\quad S_n(f, x) - f(x) = \dfrac{2^{-\alpha-\beta-1}}{\Gamma(\alpha+1)} \dfrac{\Gamma(n+\alpha+\beta+2)}{\Gamma(n+\beta+1)}$

$\qquad\qquad \cdot \int_0^\pi [T_\psi f(x) - f(x)] \, P_n^{(\alpha+1,\,\beta)}(\cos s\,\psi) \, w(\cos s\,\psi) \, (\sin\psi) \, d\psi$

(see Yadav [9] page 40, equation (2.8)) where T_ψ is the generalized translate of f in $[-1, +1]$ as explained in (4.1.8). Also, by the orthogonal property of Jacobi polynomials

(4.3.22) $\quad S_n(f, 1) - A = \dfrac{2^{-\alpha-\beta-1}}{\Gamma(\alpha+1)} \dfrac{\Gamma(n+\alpha+\beta+2)}{\Gamma(n+\beta+1)}$

$\qquad\qquad \cdot \int_0^\pi \{f(\cos s\,\psi) - A\} P_n^{(\alpha+1,\,\beta)}(\cos s\,\psi) \, w(\cos s\,\psi) \, (\sin\psi) \, d\psi$

at $x = +1$ for $\alpha \geq \beta \geq -\tfrac{1}{2}$. Where w is the weight function. But from (4.1.20), we have

$$\|T_\varphi f - f\|_x \leq C \, \|f - A\|_x$$

which in turn implies

(4.3.23) $\qquad \|S_n(f, x) - f(x)\|_x \leq A \|S_n(f, 1) - A\|_x$

But the right hand side tends to zero as the F-J series converges to A at $x = +1$ under the present conditions of the Theorem 4.3.4 for $\alpha = \beta = -\tfrac{1}{2}$ (see Theo. 2.4.7). So one gets

(4.3.24) $\qquad \|S_n(f, x) - f(x)\|_x \to 0$, as $n \to \infty$.

This completes the proof of Theorem 4.3.4.

4.4 Approximation through absolute summabilities

We recollect the idea that convergence of any transformed sequence of F-J series at $x = +1$ leads approximation of the signal by Jacobi polynomials. Many such convergence problems are given in section §3.2 where approximations are to be investigated. An absolute Nörlund summability of F-J series has been proved in Yadav[1] at $x = +1$. Again in a later investigation (Yadav[23]), it is proved that the absolute Nörlund summability holds for some unbounded signals. This is crucial to note that as a consequence, some unbounded signals can be represented by a linear sum of Jacobi polynomials.

Let $\{p_n\}$ be a sequence of non-negative real numbers such that

(4.4.1) $P_n = p_0 + p_1 + p_2 + \dots + p_n \neq 0$

where the sequence $\{p_v - p_{v+1}\}$ is non-negative and non-increasing. Again, let $\{m_v\}$ be a sequence of real numbers such that

(4.4. 2) $$\sum_{v=2}^{\infty} \frac{m_v \log v}{P_v} < \infty.$$

Then the Nörlund sum (see Nörlund[1]) $N_n(f, x)$ of factored F-J series defined by

(4.4.3) $N_n(f, x) \equiv N_n(f, x, X_N^{\alpha, \beta}) = \frac{1}{P_n} \sum_{v=0}^{n} p_{n-v} S_v(f, x)$

where $S_v(f, x)$ is partial sum of factored F-J series $\sum m_v \, a_v \, P_v^{(\alpha, \beta)}(x)$. On this preparation, following Theorem is proved in Yadav[23].

THEOREM   4.4.1   *Let the signals of the subspace $X_{Np}^{\alpha, \beta} \subset X$ satisfy the "pole" condition*

(4.4.4) $\int_0^t |F(\varphi)| \, d\varphi = o(t^{2\alpha+2} \log 1/t)$, *as* $t \to 0+$

where

(4.4.5) $F(\varphi) = \{f(\cos \varphi) - A\} (\sin \varphi/2)^{2\alpha+1} (\cos \varphi/2)^{2\beta+1}$

A be an absolute constant depending upon f only and the 'antipole' condition

(4.4.6) $\int_0^h \varphi^{\beta-\alpha} |f(-\cos\varphi)| \, d\varphi = O(\log 1/h)$, *as* $h \to 0$

holds. Let 'Lebesgue Point Like' condition that the integral:

(4.4.7) $\left| \frac{1}{t^{2\alpha+1}} \int_\omega^{\omega+t} d F(u) \right| \leq O(\log 1/t)$ *as* $t \to 0+$

holds for almost all ω in $0 < \omega < \pi$. If $\{p_n\}$ and $\{m_v\}$ satisfies (4.4.1) and (4.4.2). Then the signals $f \in X_{Np}^{\alpha, \beta}$ can be represented by the linear combinations of Jacobi polynomials $N_n(f, x)$ to any degree of accuracy.

Or in other words

(4.4.8) $$\|N_n(f, x) - f(x)\|_{X_{Np}^{\alpha,\beta}} \to 0, \ as \ n \to \infty.$$

Following Lemmas are used in the proof of the Theorem.

LEMMA 4.4.2 (Yadav[1]) *Let* $\sum u_n(x)$ *be an infinite series, such that*

(4.4.9) $$B_n = \sum_{v=0}^{n} v \, u_v(x) = O(n \log n)$$

and the sequences $\{p_v\}$ *and* $\{m_v\}$ *satisfies* (4.4.1) *and* (4.4.2) *respectively. Then the series* $\sum m_n u_n(x)$ *is absolutely Nörlund summable or summable* $|N, \ p_n|$ *at the point* x.

LEMMA 4.4.3 *If* $\frac{1}{2} \geq \alpha, \beta \geq -\frac{1}{2}$ *and the signal* $f \in X_{Np}^{\alpha,\beta}$. *Then at* $x = +1$,

(4.4.10) $$S_n(x) = \sum_{v=0}^{n} (v+1) a_v P_v^{(\alpha, \beta)}(x) = O(n \log n)$$

Moreover, the F-J series with factor m_v *i.e* $\sum_{v=0}^{\infty} m_v \, a_v P_v^{(\alpha, \beta)}(1)$ *is absolutely Nörlund summable* $(N, \ p_n)$, *where* $\{p_n\}$ *and* $\{m_v\}$ *satisfies* (4.4.1) *and* (4.4.2) *respectively. As a consequence the sequence of Nörlund mean*

(4.4.11) $$\{t_n\} \equiv \left\{ \sum_{v=0}^{n} \frac{p_{n-v}}{p_n} \sum_{i=0}^{v} m_i \, a_i P_i^{(\alpha, \beta)}(x) \right\}$$

is convergent at $x = +1$.

Proof : The last assertion in the Lemma is an outcome of the argument that the absolute summability implies the ordinary summability. Hence convergence of (4.4.11) is approved. Again the absolute Nörlund summability of $\sum_{v=0}^{\infty} m_v \, a_v P_v^{(\alpha, \beta)}(1)$ is assured by lemma 4.4.2 proved in Yadav[1] provided relation (4.4.10) holds. This is done in a long process of calculation. The partial sum of (4.4.10) at $x = 1$ is written as

(4.4.12) $$S_n(1) - A = \sum_{v=0}^{n} (v+1) a_v P_v^{(\alpha, \beta)}(1) - A$$

$$= \int_0^{\pi} F(\varphi) S_n(1, \cos\varphi) \, d\varphi = \int_0^{\pi} F(\varphi) \{S_n^1(1, \cos\varphi) + S_n^2(1, \cos\varphi)\} d\varphi$$

(for an absolute constant A depending upon the signal f by the use of Abel's transformation, the summation formula (1.2 .2) and the orthogonal property of Jacobi polynomials) where

(4.4.13) $$S_n^1(1, \cos\varphi) = \sum_{v=0}^{n-1} \frac{2^{-\alpha-\beta-1}}{\Gamma(\alpha+1)} \frac{\Gamma(v+\alpha+\beta+2)}{\Gamma(v+\beta+1)} P_v^{(\alpha+1, \beta)}(\cos\varphi)$$

and

(4.4.14) $$S_n^2(1, \cos\varphi) = (n+1) \frac{2^{-\alpha-\beta-1}}{\Gamma(\alpha+1)} \frac{\Gamma(n+\alpha+\beta+2)}{\Gamma(n+\beta+1)} P_n^{(\alpha+1, \beta)}(\cos\varphi).$$

To estimate the order of (4.4.13), the interval of integration is broken into three parts as:

$$\int_0^\pi \;=\; \int_0^{c/n} \;+\; \int_{c/n}^{\pi-c/n} \;+\; \int_{\pi-c/n}^\pi \;=\; J_1 + J_2 + J_3 \text{ (say)}.$$

Now, applying the 'pole' and the 'antipole' conditions, it is straight that

(4.4.15) $\qquad\qquad\qquad J_1, J_3 \;=\; O(n \log n)$.

To estimate J_2, one uses the inequality (1.3.6) so that

$$J_2 \;=\; \int_{c/n}^{\pi-c/n} F(\varphi) \left\{ \sum_{\nu=0}^{n-1} \nu^{\alpha+1/2} A(\nu) \, (\sin \varphi/2)^{-\alpha-3/2} (\cos \varphi/2)^{-\beta-1/2} \right\}$$

$$\times \left[\left\{ \cos\left(\nu + \frac{\alpha+\beta}{2} + 1\right)\varphi - \frac{\left(\alpha+\frac{1}{2}\right)\pi}{2} \right\} + (\nu \sin\varphi)^{-1} O(1) \right] d\varphi$$

where $A(\nu)$ is function of $\nu = 0, 1, 2, \ldots, \pi, \alpha$ and β but uniformly bounded for all of its variables. In the first sum, we use the order by McFadden[1]

(4.4.16) $\qquad \sum_{\nu=0}^{n-1} \nu^{\alpha+1/2} A(\nu) \, e^{i\left(\nu+\frac{\alpha+\beta}{2}+1\right)\varphi} \;=\; O\left(n^{\alpha+\frac{1}{2}} \varphi^{-1}\right)$

for $-\frac{1}{2} \leq \alpha \leq \frac{1}{2}$ to get the desired order and a direct calculation of second sum is $O(n \log n)$. Similar procedure is adopted for the second part $S_n^2(1, \cos\varphi)$ and finally we reach at the order

(4.4.17) $\qquad\qquad\qquad J_2 = O(n \log n)$.

This finalizes the proof of (4.4.10) and finishes the proof of the Lemma.

Proof of the Theorem 4.4.1 Given that $f \in X_{Np}^{\alpha,\beta}$ and satisfies (4.4.4) and (4.4.6), so that by Lemma 4.4.3 the factored F-J series with factor $\{m_\nu\}$ is absolutely Nörlund summable,, consequently the linear combination $N_n(f, x)$ of Jacobi polynomials converges at $x = +1$. But from (4.1.20), for $\alpha \geq \beta \geq -\frac{1}{2}$,

(4.4.18) $\qquad\qquad \|T_\psi f - f\|_X \leq C \|f - A\|_X$ (C an absolute constant),

which after the argument in (4.3.15) gives

(4.4.19) $\qquad\qquad \|N_n(f, x) - f(x)\|_{X_{Np}^{\alpha,\beta}} \to 0, \; as \; n \to \infty,$

when $+\frac{1}{2} \geq \alpha \geq \beta \geq -\frac{1}{2}$. This completes the proof of the Theorem 4.4.1.

REMARK 4.4.2 Absolute summability at $x = +1$ of the F-J series has been discussed in detail in the section §3.2 of Chapter III. Again many more cases of absolute summabilities of F-J series are left to be investigated. It indicates that a vast area is open to furthering the advancement in

Information Technology which is expected to revolutionize the coming days vision. The technique developed is applicable to the non-manageable signals in the IT, as we are now, in a position to transfer an infinite signal into a linear combination of Jacobi polynomials.

CHAPTER V

SIGNAL PROCESSING AND WAVELETS

5.1 Introduction and definitions

The word 'wavelet' is popular among scientists linked with different fields like Mathematics, Physics, Engineering, Bioscience and Oil and natural Gases etc. Wavelet analysis has begun to innovate scientific thinking in a broad range of applications including Signal processing, Error corrections in Data and Image processing, Solutions of Partial Differential Equations, Modeling Multiscale Phenomena and Statistics (Resnikoff and Wells [1], Igari [1]). We have picked up some ideas to use wavelets in Signal Processing mainly to represent a bunch of Signals in terms of Coifman wavelet systems.

Definition 5.1.1 **Coifman Wavelet Systems** *An orthonormal wavelet system with compact support is called a Coifman wavelet system of degree N if the moments of φ and ψ satisfy the relations:*

(5.1.1) $\text{Mom } (\varphi)_l = \int \varphi(x)dx = 1,$

(5.1.2) $\text{Mom } (\varphi)_l = \int x^l \varphi(x)\, dx = 0, \quad (l = 1, 2, \dots N)$

(5.1.1) $\text{Mom } (\psi)_l = \int x^l \psi(x)dx = 0, \quad (l = 0, 1, 2, \dots N),$

where φ is scaling function and ψ is mother wavelet.

Notations: From now on, all functions synonymously called signals associable with F-J series are denoted by X to mean the space C[a, b] or $L_p^{\alpha,\beta}[a,b]$.   C[a, b] is Banach space of all continuous functions on [a, b] while $L_p^{\alpha,\beta}[a,b]$, $(1 \le p \le \infty)$ are Banach spaces of p-power Lebesgue integrable functions on [a, b] with weight w(x) such that

(5.1.4) $w(x) = \begin{cases} w(x), & x \in [a,b] - \infty < a < b < +\infty; \\ 0 & otherwise. \end{cases}$

α, β are indicatives of w which converts itself to w(x) of definition 1.1.1 when and where the context requires. Usual sup, p-norm and ess sup norms are defined there. Thus

$f : [a,b] \to \mathbb{R}$ with $\|f\, x^r w\|_X < \infty$, $(r = 0, 1, 2, \dots)$ so that $f \in C[a,b]$ or $\in L_p^{\alpha,\beta}[a,b]$. We assume f continuous on [a, b] for p $= \infty$. Also, by $C_0^N[a,b]$ we mean the set of all continuous functions tending to zero at infinity, differentiable N-times on [a, b] with Nth derivative continuous.

Let $\wp$ denote the linear span of all Orthogonal Polynomials dense on the Hilbert space $L_2^{\alpha,\beta}[a,b]$. Then the following proposition holds (see Szegö[1] page 40 Theorem 3.1.5).

Proposition 5.1.2 *The set of all polynomials $\wp$ is closed in $L_2^{\alpha,\beta}[a,b]$ and hence in our terminology $\wp$ is dense on $L_2^{\alpha,\beta}[a,b]$, i.e. $L_2^{\alpha,\beta}[a,b] \subseteq \overline{\wp}$, or $L_2^{\alpha,\beta}[a,b]$ is covered by the closure of $\wp$.*

Definition 5.1.3 **Approximation by polynomials**

Let $\sigma_n(x, \wp)$ be a combination of elements of $\wp$ each of which is of degree less than or equal to n so that the resulting polynomial is of degree n and belongs to $\wp$. If an $f \in X$ is such that

(5.1.5a) $\forall \varepsilon > 0, \exists n_0: n \geq n_0 \Rightarrow \|f - \sigma_n\|_X < \varepsilon$

Then it is said that f is approximable by a polynomial of degree n $(n \geq n_0)$. This in turn is said:

(5.1.5b) $\|f(x) - \sigma_n(x, \wp)\|_X \to 0$, as $n \to \infty$.

Proposition 5.1.4 *Polynomial $\sigma_n(x, \wp)$ is differentiable N –times $(N > n)$ so that its Nth derivative is continuous on $[a, b]$ and*

(5.1.6) $\sigma_n(x, \wp) \chi_{[a,b]} \in C_0^N[a,b]$

for finite a, b ; arbitrary N and $\chi_{[a,b]}$ the characteristic function of $[a, b]$.

Proof. In view of (1.1.1) we see that nth derivative of σ_n and later ones are constants thus continuous and differentiable, while $\sigma_n(x, \wp)$ $\chi_{[a,b]}$ is zero at infinity for finite a, b $\in \mathbb{R}$. Moreover, N is arbitrary non-negative integer in (5.1.6). Hence we say that σ_n is differentiable arbitrarily.

Proposition 5.1.5 *For $f, g \in X$ and $f, g \in L_2^{\alpha,\beta}[a, b]$ there exist absolute constant M such that for any scalars α, β*

(5.1.7) $\|\alpha f + \beta g\|_X \cong \|\alpha f + \beta g\|_{L_2^{\alpha,\beta}}$

Proof. The result in the proposition 5.1.5 is an outcome of the fact that the norms in X and $L_2^{\alpha,\beta}$ are finite real numbers and $f, g \in X \cap L_2^{\alpha,\beta}[a, b]$. This completes the proof.

Definition 5.1.6 **Wavelet sampling approximation operator**

An operator S^j defined on $C_0^N[\mathbb{R}]$ and given by

(5.1.8a) $S^j(f)(x) = 2^{-j/2} \sum_{k \in z} f\left(\frac{k}{2^j}\right) \varphi_{jk}(x)$

where $\{\varphi_{jk}\}$ is Coifman wavelet system such that

(5.1.8b) $\varphi_{jk}(x) = m^{\frac{j}{2}} \varphi(m^j x - k), \quad k, j \in Z$

(m a fixed positive integer related to the rank of the wavelet matrix) forms jth level basis, is called Wavelet Sampling Approximation Operator, (in short **Sampling Operator**).

For details of these Operators, one may see the treatise *'Wavelet Analysis'* by Resnikoff and Wells[1].

Definition **5.1.7** **Wavelet orthogonal projection operator**

The operator P^j given by

(5.1.9) $P^j(f)(x) = \sum_{k \in Z} \left(\int_R f(t)\, \varphi_{j\,k}(t)\, w(t)dt \right) \varphi_{j\,k}(x)$

where $\varphi_{j\,k}$ is as above; is called Wavelet Orthogonal Projection Operator, (in short **Projection Operator**).

Recently Tian and Wells[1] have proved the following theorems for $C_0^N[\mathbb{R}]$ class of functions. This provides a clue for reconstructing a signal by given samples of continuous strips.

THEOREM   5.1.8   *For an Orthonormal Coifman wavelet system of degree N with the scaling function $\varphi(x)$ and scaling vector α_1, assume α_1 has finite length. If $f \in C_0^N[\mathbb{R}]$, then*

(5.1.10) $\left\| f(x) - S^j(f)(x) \right\|_{L^2} \leq C_1 2^{-jN}$

is jth level approximation of $f \in L^2[\mathbb{R}]$, where C_1 is a constant which depends upon f and the scaling vector α_1.

THEOREM   5.1.9   *Under the same hypothesis of Theorem 5.1.8, we have*

(5.1.11) $\left\| f(x) - P^j(f)(x) \right\|_{L^2} \leq C_2 2^{-jN}$

where C_2 is a constant which depends upon f and the scaling vector α_1.

5.2   Lebesgue class and wavelet approximation

Theorems 5.1.8 and 5.1.9 are tools of my creations. These provides clue to transfer a signal of class $C_0^N[\mathbb{R}]$ into various levels of approximation by wavelets. We have observed that a Lebesgue integrable signal can be transferred into wavelet form. Our observation is based on Approximation by Fourier-Jacobi Expansions, and found that it may prove to be a powerful technique in Information Technology in developing algorithms for *softwares* useful in contaminated noise removal, corrupted data repairs, lost data recovery, compression of data volume, data understanding, visualization and security etc. Further advancement in this direction may be writing of generalized functions in terms of wavelets. Our general results   proved in Yadav [20] is as follows.

THEOREM   5.2.1   *Let X be a subspace of $L^1[a, b]$ functions such that the approximation in (5.1.5) holds. Then for every $f \in X$, there exists an integer n_0 such that*

(5.2.1) $\left\| f(x) - S^j\left(\sigma_{n_0} \chi_{[a,b]} \right)(x) \right\|_X \leq C_3 2^{-jN}$

where C_3 is a constant may be depending upon f and the scaling vector α_1. N is any non-negative integer associated to the degree of Coifman wavelet systems. S^j is jth level sampling operator given by (5.1.8).

THEOREM 5.2.2   *Under the same hypotheses of Theorem 5.2.1, $\forall f \in X$ there exists an integer n_0 such that*

$$(5.2.2) \qquad \left\| f(x) - P^j\left(\sigma_{n_0}\chi_{[a,b]}\right)(x) \right\|_X \leq C_4 2^{-jN}$$

where jth level Projection Operator P^j is given by (5.1.9). The constant C_4 may depend upon f and the scaling vector α_1. N is the degree of Coifman wavelet systems.

Proof of Theorem 5.2.1   Since ε is arbitrary in (5.1.5a), so we have

$$(5.2.3) \qquad \left\| f - \sigma_{n_0} \right\|_X < 2^{-jN} \implies \left\| f - \sigma_{n_0}\chi_{[a,b]} \right\|_X < 2^{-jN}$$

as we are considering approximation in [a, b]. Also by definitions

$$S^j\left(\sigma_{n_0}\chi_{[a,b]}\right)(x) = 2^{-j/2}\sum_{k \in z}\left(\sigma_{n_0}\chi_{[a,b]}\right)\left(\frac{k}{2^j}\right)\varphi_{j\,k}(x).$$

Thus

$$\left\| f(x) - S^j\left(\sigma_{n_0}\chi_{[a,b]}\right)(x) \right\|_X$$

$$\leq \left\| f(x) - \sigma_{n_0}(x) \right\|_X + \left\| \sigma_{n_0}(x) - S^j\left(\sigma_{n_0}\chi_{[a,b]}\right)(x) \right\|_X$$

$$< 2^{-jN} + C_5 \left\| \sigma_{n_0}(x) - S^j\left(\sigma_{n_0}\chi_{[a,b]}\right)(x) \right\|_{L^2}$$

$$\leq C_6 2^{-jN}\qquad .$$

For polynomials $\sigma_{n_0}(x) \in C_0^N[\mathbb{R}]$ are square integrable. This completes the proof of Theorem 5.2.1.

Proof of Theorem 5.2.2   We have,

$$P^j\left(\sigma_{n_0}\chi_{[a,b]}\right)(x) = \sum_{k \in z}\left(\int_R\left(\sigma_{n_0}\chi_{[a,b]}\right)(t)\,\varphi_{j\,k}(t)\,w(t)dt\right)\varphi_{j\,k}(x)$$

$$= \sum_{k \in z}\left(\int_a^b\left(\sigma_{n_0}\right)(t)\,\varphi_{j\,k}(t)\,w(t)dt\right)\varphi_{j\,k}(x)$$

Now

$$\left\| f(x) - P^j\left(\sigma_{n_0}\chi_{[a,b]}\right)(x) \right\|_X \leq \left\| f(x) - \sigma_{n_0}(x) \right\|_X$$

$$+ \left\| \sigma_{n_0}(x) - P^j\left(\sigma_{n_0}\chi_{[a,b]}\right)(x) \right\|_X$$

$$< 2^{-jN} + C_7 \left\| \sigma_{n_0}(x) - p^j\left(\sigma_{n_0}\chi_{[a,b]}\right)(x) \right\|_{L^2}$$

(for, polynomials σ_{n_0} and $\varphi_{jk}(x)$ are L^2 functions.)

$$\leq\ C_6\ 2^{-jN}.$$

This completes the proof of Theorem 5.2.1.

From the proofs of Theorems 5.2.1 and 5.2.2, it is clear that an approximation of a signal by polynomials allows one to write the signal in the form of wavelet series. This provides to have (i) *Efficient Algorithms* for representing signals in terms of wavelet basis (ii) *Compression Algorithms* for archiving, Computation and Information and (iii) *Channel Coding* for efficient way for transmission over noisy channels, etc. We have proved many approximations under different circumstances. Some theorems related to wavelet representation are being given as proved in Yadav[17] while many more are left to be done in coming days. To deal with Jacobi polynomials we now work with the interval $[-1, +1]$ instead of [a, b] and weight $w(x) = (1 - x)^\alpha (1 + x)^\beta$, $(\alpha, \beta > -1)$ as given in the Definition 1.1.1.

THEOREM   5.2.3   *Let $X_1^{\alpha,\beta} \subseteq X$ be a subspace of signals which satisfy the 'Pole' condition,*

(5.2.4) $\int_0^t |f(\cos\varphi) - A|\ \varphi^{2\alpha+1}\ d\varphi\ =\ o(t^{\,2\alpha+2})$, *as* t $\to$ 0+

for $+1/2 > \alpha \geq \beta \geq -\tfrac{1}{2}$ (but for $\beta > -\tfrac{1}{2}$, $\alpha + \beta \leq 0$) where A is an absolute constant depending upon f. Then the approximation

(5.2.5) $\|\sigma_n^1 (f, x) - f(x)\|_X \to 0$, *as* n $\to$ ∞;

holds, where the linear combination of Jacobi polynomials

(5.2.6) $\sigma_n^1 (f, x) \equiv \frac{1}{n+1}\sum_{k=0}^{n} s_n (f, x)$

is Cesàro mean of order one of the Jacobi series (2.1.8) with s_n its partial sum. Moreover, for every $f \in X_1^{\alpha,\beta}$ there exists a natural number n_0 such that for $n \geq n_0$, the wavelet representation,

(5.2.7) $\|f(x) - S^j(\sigma_n^1 \chi_{[-1, 1]})(x)\|_X \leq C_8 2^{-jN}$

and

(5.2.8) $\|f(x) - P^j(\sigma_n^1 \chi_{[-1, 1]})(x)\|_X \leq C_9 2^{-jN}$,

hold, where the operators S^j and P^j are Sampling and Projection operators respectively.

Proof of the Theorem   5.2.3   The approximation (5.2.5) of f in terms of Jacobi polynomials is proved in Yadav[16] as shown in (4.2.1). The results (5.2.7) and (5.2.8) are obtained by the applications of Theorems 5.2.1 and 5.2.2 respectively. This completes the proof of the Theorem.

Following Theorems proved in Yadav[19] gives wavelet representation for signals of another subspace $X_2^{\alpha,\beta} \subseteq X$. Signals of the subspace $X_2^{\alpha,\beta}$ satisfy the antipole condition: that the integral

(5.2.9) $\qquad h^{\frac{1}{2} - \alpha} \int_0^h |f(- \cos \varphi)| \, \varphi^{\beta + 1/2} \, d\varphi = o(1),$ as $h \to 0+$

exists, in addition to the pole condition (5.2.4). The Theorem reads:

THEOREM 5.2.4 *Let the signals of the subspace $X_2^{\alpha,\beta} \subseteq X$ satisfy the pole condition (5.2.4) and the antipole condition (5.2.9) for $\frac{1}{2} > \alpha \geq \beta \geq -\frac{1}{2}$ (where for $\beta > -\frac{1}{2}$ one has $\alpha + \beta > 0$). For $\beta \geq -\frac{1}{2}$ but $\alpha + \beta \leq 0$, no antipole condition is necessary. Then the approximation of an $f \in X_2^{\alpha,\beta}$ in terms of a linear combination of Jacobi polynomials given by (5.2.6) i.e.*

(5.2.10) $\qquad \left\| \sigma_n^1 (f, x) - f(x) \right\|_X \to 0,$ as $n \to \infty;$

holds and there exists a natural number n_0 such that for all $n \geq n_0$ the wavelet representation

(5.2.11) $\qquad \left\| f(.) - S^j\left(\sigma_n^1 \chi_{[-1, 1]}\right)(.) \right\|_X \leq C_{10} 2^{-jN}$

and

(5.2.7) $\qquad \left\| f(.) - P^j\left(\sigma_n^1 \chi_{[-1, 1]}\right)(.) \right\|_X \leq C_{11} 2^{-jN}$

hold. The operators S^j and P^j are defined by (5.1.8) and (5.1.9) respectively.

Proof of Theorem 5.2.4 Proof of (5.2.10) is given in Yadav[19] presented here in this monograph as Theorem 4.2.2 in chapter IV. Proofs of (5.2.11) and (5.2.12) are consequences of the relation (5.2.10) and arguments are completed as the proofs of Theorems 5.2.1 and 5.2.2. This completes the proof of the Theorem 5.2.4.

Following particular case solved in Yadav[22] is also interesting.

THEOREM 5.2.5 *Let $X_6^{\alpha,\beta} \subset X$, $(\alpha = \beta = -\frac{1}{2})$ be a subspace of signals satisfying the 'pole' condition*

(5.2.13) $\qquad \int_0^t |f(\cos \varphi) - A| \, \varphi^{\alpha - 1/2} \, d\varphi = o(t^{\alpha + 1/2}),$ as $t \to 0+$

Then f can be represented by the partial sum $S_n(f, x)$ of F-J series to any degree of accuracy. Moreover, for every Coifman wavelet system of degree N, there exists natural numbers n_0 such that the wavelet representation

(5.2.14) $\qquad \left\| f(.) - S^j\left(S_{n_0} \chi_{[-1, 1]}\right)(.) \right\|_X \leq C_1 2^{-jN}$

and

(5.2.15) $\qquad \left\| f(.) - P^j\left(S_{n_0} \chi_{[-1, 1]}\right)(.) \right\|_X \leq C_2 2^{-jN}$

holds for $n > n_0$. S^j, P^j are Sampling and Projection operators defined by (5.1.8) and (5.1.9) and C_i ($i = 1, 2, ...$) are absolute constants not the same at each occurrence.

Proof of the Theorem 5.2.5 Approximation of an $f \in X_6^{\alpha,\beta}$ by the sequence of the partial sums { $S_n(f, x)$ } of Jacobi series for $\alpha = \beta = -\frac{1}{2}$ has been derived as Theorem 4.3.5. The conclusions

(5.2.14) and (5.2.15) are obtained on the lines of the proofs of (5.2.1) and (5.2.2) respectively. This completes the proof of Theorem 5.2.5.

Question that what type of Lebesgue integrable signals can be transformed into a wavelet series has always been a challenge to many adventurers doing something new in *Information Technology*. Various types of signals, Lebesgue integrable and satisfying certain additional conditions, were approximated by some linear combinations of Jacobi polynomials in Yadav [18] and then by wavelet series in Yadav[17]. The best possible circumstance has been discovered. We explain the same here to draw a boundary beyond which signals are not representable by the Cesàro method of summation but one has to wander for some other methods, indicated in the last section of Chapter IV of this monograph. For this purpose, we mention some notations. By σ_n^k we mean the linear combination of Jacobi polynomials which is Cesàro mean of order k of Jacobi series (2.1.8) given as

$$(5.2.16) \qquad \sigma_n^k\,(f,\,x) \equiv \sigma_n^k\left(f,\,x,\,X_p^{\alpha,\beta}\right)$$

$$= \sum_{v=0}^n \frac{A_{n-v}^k}{A_n^k}\,\hat{f}(v)\omega_v^{(\alpha,\beta)}\,R_v^{(\alpha,\beta)}(x)$$

$$= \sum_{v=0}^n \frac{A_{n-v}^{k-1}}{A_n^k}\,s_v(f,x).$$

where s_v is the vth partial sum of the F-J series. A_i^k, ($i = 0, 1, 2, ...$) are bionomial coefficients of x^i in the expansion of $(1-x)^{-k-1}$ (see eqn. (5) of §4.3). Now a subspace $X_k^{\alpha,\beta}$ is considered where a pole condition

$$(5.2.17) \qquad \int_0^t |f(\cos\varphi) - A|\,\varphi^{2\alpha+1}\,d\varphi = o(t^{2\alpha+2}),\ \text{ as }\ t \to 0+$$

is lighter than the continuity of f at $x = \cos\varphi = +1$. Moreover, the antipole condition

$$(5.2.18) \qquad \int_0^h |f(-\cos\varphi)|\,\varphi^{\beta-\alpha+k}\,d\varphi = o(1),\ \text{as } h \to 0+$$

is weaker than that in Szegö ([1] Theorem 9.1.4) which were explained in Remark 2.3.9 . Following Theorem is being proved with extreme boundary.

THEOREM 5.2.6   *The approximation of a signal* $f \in X_k^{\alpha,\beta} \subset X$ *satisfying the pole condition (5.2.17) and the antipole condition (5.2.18) for* $\alpha \geq \beta \geq -\frac{1}{2}$, *by* $\sigma_n^k\,(f,\,x)$ *holds. For the wavelet representations, there exists natural number* n_0 *depending upon the degree of Coifman wavelet systems so that the approximations*

$$(5.2.19) \qquad \left\|f(.) - S^j\big(\sigma_{n_0}^k \chi_{[-1,\,1]}\big)(.)\right\|_X \leq C_1 2^{-jN}$$

and

$$(5.2.20) \qquad \left\|f(.) - P^j\big(\sigma_{n_0}^k \chi_{[-1,\,1]}\big)(.)\right\|_X \leq C_2 2^{-jN}$$

are valid for $\alpha \geq \beta \geq -\frac{1}{2}$, *provided the pole condition (5.2.17) holds for* $k \geq \alpha + \beta + 1$. *But for* $\beta > -\frac{1}{2}$, $\alpha + \frac{1}{2} < k < \alpha + \beta + 1$, *the additional 'antipole' condition (5.2.18) is required to be satisfied by f* $\in X_k^{\alpha,\beta}$. *For* $-1 < \beta \leq -\frac{1}{2}$ *or (for* $\beta > -\frac{1}{2}$, *but* $k \geq \alpha + \beta + 1$ *) no antipole condition is necessary. For* $k \leq \alpha + \frac{1}{2}$ *or for* $\beta > -\frac{1}{2}$, *and* $\alpha + \frac{1}{2} < k < \alpha + \beta + 1$ *but without the antipole condition; the statement in not correct.*

Proof of Theorem 5.2.6   Approximation of an $f \in X_k^{\alpha,\beta}$ under the same conditions has been stated and solved in Theorem 4.3.1. Again the wavelet representation using operators S^j and P^j are outcomes of the methods applied in the proofs of (5.2.1) and (5.2.2) respectively. This finishes the proof of the Theorem.

5.3   Application of absolute convergence in wavelet representation

This is interesting to note that how pure mathematics is in the root of applied one. Absolute convergence of an infinite series at a point or in an interval implies its ordinary convergence at the point or in the interval. This concept is playing vital role in our applied mathematical formula given below. In a big jump, we state and prove the wavelet representations for signals of the subspace $X_{Np}^{\alpha,\beta} \subset X$ which is attributed by the conditions (4.4.4), (4.4.6) and (4.4.7). For clear vision, we repeat them here. The signals $f \in X_{Np}^{\alpha,\beta} \subset X$ satisfy the "pole" condition

$$(5.3.1) \qquad \int_0^t |F(\varphi)| \ d\varphi \ = \ o(t^{2\alpha+2} \ log \ 1/t), \text{ as } \ t \to 0+$$

where

$$F(\varphi) \ = \ \{f(cos \ \varphi) - A\} \ (sin \ \varphi/2)^{2\alpha+1} \ (cos \ \varphi/2)^{2\beta+1}.$$

A is an absolute constant depending upon *f* only and the 'antipole' condition

$$(5.3.2) \qquad \int_0^h \varphi^{\beta-\alpha} |f(-cos \ \varphi)| \ d\varphi \ = \ O(log \ 1/h \ ), \text{ as } h \ \to \ 0+;$$

holds. Also, a *'Lebesgue Point Like'* condition that the integral:

$$(5.3.3) \qquad \left| \frac{1}{t^{2\alpha+1}} \int_\omega^{\omega+t} d \ F(u) \right| \ \leq \ O(log \ 1/t \ ) \text{ as } t \ \to 0+$$

holds for almost all ω in $0 < \omega < \pi$. Again the real sequences $\{p_v\}$ and $\{m_v\}$ satisfy (4.4.1) and (4.4.2) of section §4.4 with $\alpha \geq \beta \geq -\frac{1}{2}$. Then it is proved that *f* is represented by the linear combination of Jacobi polynomials

$$(5.3.4) \qquad N_n(f,x) \equiv N_n(f,x,X_{Np}^{\alpha,\beta}) = \frac{1}{P_n} \sum_{v=0}^n p_{n-v} \ S_v(f,x)$$

$$= \frac{1}{P_n} \sum_{v=0}^n p_{n-v} \sum_{i=0}^v m_i \ a_i \ P_i^{(\alpha, \ \beta)}(x)$$

For vivid description one may see the section §4.4. These conditions themselves explain the diverging behavior of signals of the subspace $X_{Np}^{\alpha,\beta}$ whose elements are being represented by wavelets. Then our state of the art Theorem is

THEOREM   5.3.1   *Let $f \in X_{Np}^{\alpha,\beta} \subset X$ satisfy the conditions of Theorems 4.4.1. Then f is approximable by the operator $N_n(f,x)$ on [- 1, +1] and there exists natural number n_0 depending upon the degree of the Coifman wavelet systems such that the wavelet representations*

(5.3.5) $$\|f(.) - S^j(N_n\chi_{[-1,\ 1]})(.)\|_X \leq C_2 2^{-jN}$$

and

(5.3.6) $$\|f(.) - P^j(N_n\chi_{[-1,\ 1]})(.)\|_X \leq C_2 2^{-jN}$$

hold for $\alpha \geq \beta \geq -\frac{1}{2}$ whenever $n \geq n_0$. S^j and P^j are Sampling and Projection operators. $\chi_{[-1,\ 1]}$ is characteristic function of [- 1, +1].

Proof of the Theorem   5.3.1   Representation of f by $N_n(f,x,X_{Np}^{\alpha,\beta})$ has been proved in Theorem 4.4.1 under the same conditions as given here. Then the Sampling and Projection operators are applied to get f in terms of wavelets as done in (5.2.1) and (5.2.2). This completes the proof of Theorem 5.3.1.

The subject '*Approximation of a signal by polynomials then by wavelets*' will always remain interesting. It will play important roles in Mathematical Analysis for investigations and applications in various fields related to human progress. I hope this monograph will get hot welcome by interested researchers and mathematicians. As the subject of Absolute or Ordinary summabilities of F-J series at $x = +1$ advances so will advance the techniques of representing a signal in terms of wavelets. Best approximation can also be utilized for polynomial and wavelet representation of a signal (see Yaadav[15]). Many more bunches of non-manageable signals can be represented by wavelets if the Theorems on absolute summabilities already derived and to be derived as mentioned in section §3.2 are used. Moreover, this opens a vast area of interesting researches where useful results are waiting. Interested researchers are requested to be careful for misprints and slips.

I have lived my life with this neglected Shoppe of mathematics with the hope that a time will come when some great hearted persons will rejoice for a short while with these writings.

REFERENCES

Adamoff, A.

1.   *Expansion of an Arbitrary Function of a Single Real Variable in Series of Functions of a Preassigned kind* (in Russian). Thesis St. Petersburg, 1907, 191 pp.

Askey, R. and Wainger, S.

1.   *A convolution structure for Jacobi polynomials.* Amer. J. Math 91 (1969) 463 - 485.

Bavinck, H.

1.   *Approximation Processes for Fourier-Jacobi Expansion.* Appl. Anal. 5 (1976) 296 - 312.

2.   *A Special Class of Jacobi Series and some Applications.* J. Math. Anal. Appl. 37 (1972), 767 - 797.

Butzer, P.L. and Nessel, R.J.

1.   *Fourier Analysis and Approximation.* Bd. **40**, Vol.1, Birkhauser Verlag basel unt Stuttgrat (1971).

Fejér, L.

1.   *Über die Summabilitat der Laplaceschen Reihe durch Arithmetische Mittel.* Mathematische Zeitschrift, vol 24 (1925) pp. 267 - 284.

Gasper, G.

1.   *Positivity and the Convolution Structure for Jacobi Series.* Ann. of Math. (2) **93** (1971) 112 - 116

Gupta, D.P.

1.   *Degree of convergence of Jacobi series.* Acta. Math. Acad. Scientiarum Hungaricae. 17 (1966) 15-22.

2.   *An Order Estimate for the Partial Sums of Jacobi Series.* Indian Journal of Mathematics. 9 (1967) 375-380.

3.   *On a local property of absolute summability for expansion of Fourier - Jacobi class.* Jorn. of London Math. Soc. (2) 4 (1971) 337-345.

4.   *D.Sc. Thesis*, University of Allahabad (1970).

Gupta, D. P. and Chaudhary, R. S.

1.   *Harmonic Summation of Jacobi Series.* Rend. Sc. Fis Mat. e Nat. LVIII (1975), 115-120.

Haar, A.

1.   *Reihenentwicklungen nach Legendreschen Polynomen.* Mathematische Annalen, vol.78 (1917) pp. 121 - 136.

Igari, Satoru

1.   *Real Analysis with an Introduction to Wavelet Theory.* Vol. 177, American Math. Soc. Providence, Rhode Island (1998).

Kogbetliantz, E.

1.   *Sur le Développements de Jacobi.* Comptes Rendus de l'Academie des Sciences, Paris, vol. 168 (1919) 992 – 994.

2.   *Sur les Développements de Jacobi.* Comptes Rendus de l'Academie des Sciences, Paris, vol. 172 (1921), pp. 1333 - 1334; vol. 192 (1931) pp. 915 - 918

Lukács, F.

1.   *Über die Laplacesche Reihe.* Mathematische Zeitschrift, vol. 14 (1922) pp 250 - 262.

Mc Fadden, L.

1.   *Absolute Nörlund Summability.* Duke Math. J. 9 (1942) 168 - 207.

Nörlund, N.E.

1.   *Sur   une   application   des   fonctions   permutable.* Lunds   Universitetes Arskrift. (2) 16   (1919) no. 3.

Obrechkoff, N.

1.   *Sur la Sommation de la Serie Ultraspherique, par la Méthode des Moyennes Arithmetiques.* Rendconti del Circolo Matematico di Palermo, vol. 59 (1936), pp. 266-287.

2.   *Formules asymptotiques pour les polynomes de Jacobi et sur les séries suivant les mêmes polynomes.* Annuaire de l'université de Sofia. Faculté Physico-Methematique, vol. 1 (1936), pp. 39-133.

Pandey, G.S.

1.   *On the Cesàro summability of Jacobi series.* Indian Journal of Mathematics. vol. 10 (2) (1968) 121-155.

Rau, H.

1.   *Über die Lebesgueschen Konstanten der Reihenentwicklungen nach Jacobischen Polynomen.* Journal fur die reine und angewandte Mathematik, vol. 161 (1929), pp. 234 - 254.

Resnikoff, L. and R.O. Wells, Jr.

1.   *Wavelet analysis (The scalable structure of information).* Springer-Verlag, New York Inc. (1998).

Szegö, G.

1.   *Orthogonal Polynomials*. American Math. Soc., Coll. Pub. **23** (New York) (1959).

2.   *Asymptotische Entwicklungen der Jacobischen Polynome*. Schriften der Konigs-berger Gelehrten Geselscaft, naturwissenschaftliche Klasse, vol 10 (1933) pp. 35 -112

Tian, J. and Wells, R. O.

1.   *Vanishing Moments and Biorthogonal Coifman Wavelet System*. Proc. of 4th International Conference on Mathematics in Signal Processing, University of Warwick, England 1996-1997.

Toeplitz, O.

1.   *Uber allgemeine lineare Mittelbildungen*. Prace Mathematyczno **22** (1911) 113 - 113

Whittaker, E.T. and Watson, G.N.

1.   *A course of Modern Analysis*. University Press, Cambridge; 1950.

Yadav, S. P.   (Yadav, Sarjoo Prasad)

1.   *On | N, p_n |-summability of Factored Jacobi Series at End points*. Jorn. Indian Math. Soc. **38** (1974) 329-335. 1976   **MR 52** # 14827.

2.   *On |C,1|-summability of Factored Jacobi series*. Vijnan Parishad Anusandhan Patrika (Allahabad) **21**(3) (1978) 257-260. 1981   **MR 81 b** # 42089.

3.   *The |N,p_n|-summability of factored Jacobi series at internal points*. Pure and applied Mathematika Sc, **VII(1-2)** (1978) 15-18, 1979 **MR58**# 1095.

4.   *End point absolute summability factor of Jacobi series*. Vijnana Parishad Anu. Patrika **(Alld.)** Vol. **18**(2) (1975) 101-114, 1976 **MR 52** # 3864.

5.   *On |C, 1|-summability of Jacobi series*. Ind. Jorn. Pure and Applied Math.. **8**(6) (1977) 538-45. 1978 **MR 57** # 13364.

6.   *On the Absolute Convergence of Jacobi Series at End points*. Bull. Cal. Math. Soc.**72** (1980) 87-94. **MR 82j** # 33021.

7.   *Convergence and Absolute Cesàro Summability of Jacobi Series*. The Math. Student (IMS) **47**(3) (1979) 245-51.

8.   *On Logarithmic Means of Jacobi Series*. Jorn. Ind. Math. Soc. **45** (1981) 275-83. 1987 **MR 87d** # 33028.

9.   *On the Saturation Orders of Approximation Processes Involving Jacobi Polynomials*. Jorn. Approx. Theory (USA). **51** (1) (1991) **MR 91c:** 41065.

10   *Saturation Orders of Some Approximation Processes in Certain Banach Spaces*. Studia Math. Hungarica. **28** (1993), 299-316. **MR # 96c:** 41048.

11. A Note on the Matrix Transform of Fourier-Jacobi Expansions. Ind. Jorn. Math. **34** (1992) 111-17. **MR 94i:** 41009.

12. *On the Characterization of Function Spaces in Terms of Sequential Properties.* Jorn. Ind. Math. Soc. **69** (1 - 4), (2002) 23 - 32.

13. On *the Generalized Summability Theorem of G. Szegö for Jacobi series.* Ganit Sandesh, Rajasthan Ganit Parishad. **12** (1998) 55 - 64. **MR** 2000f : 42018.

14. *Approximation by a Linear Method Involving Jacobi Polynomials.* The Math. Student Vol. 79 no. 1-4 , (2010) 193-198.

15. *On the Best Approximation by Chebyshev System of Jacobi Polynomials.* Jorn. Ramanujan Math. Soc. **17** no. **4** (2002) 261- 266. **MR** 2003k: 41A40.

16. *On a Banach Space Approximable by Jacobi Polynomials.* Acta Math. Hungar. **98** (1 - 2) (2003) 21-30. **MR** 2004a: 42039.

17. (with Rakesh Kumar Yadav)

    A Note on Wavelet Approximation. The Aligarh Bull. of Maths. vol. **22** no. **2** (2003) 161-165.

18. *On the Denseness of Jacobi Polynomials.* International Journal of Mathematics and Mathematical Sciences, (Texas, U S A). vol. **2004** (no. 28) (2004) 1455 1462.**MR** 2085014 (2005g: 41017).

19. *Some Lebesgue Subspaces Approximable by Jacobi Polynomials.* J. Indian Math. Soc. (N. S.) **74** (2007) no. 1-2, 59-69 **MR** 2422104 (2009e: 41022)

20. *Approximation of Some Lebesgue Functions by Wavelets.* J. Indian Math. Soc. vol. **73** no. 1-2 (2006), 17-23. **MR** 2290144 (2006).

21. (with Rakesh Kumar Yadav and Dinesh Kumar Yadav)

    On the Convergence of Jacobi Series at the Poles. Frontiers in Environmental Engineering (FIEE) vol **2** (2) (June 2013) wwww.seipub.org/fiee

22. (with Rakesh Kumar Yadav and Dinesh Kumar Yadav)

    Some Signals Representable by Jacobi Polynomials and Wavelets. Investigations in Mathematical Sciences vol. 3(1), 2013, 177-185.

23. (with Rakesh Kumar Yadav and Dinesh Kumar Yadav)

    On the Nörlund Method of Signal Processing Involving Coifman Wavelets. (**MIMT** Conference, Singpore (2011) IEEE 2011. Advance Materials and Research, vols. 433 - 440 (2012), pp 3378 - 3387. *Trans Tech Publications*, Switzerland.

24. *Ph. D Thesis, 'Some Problems on Summability of Jacobi Series'.* Vikram University, Ujjan. INDIA (1972)

Young, W. H.

1.　*On the Connecxion between Legendre Series and Fourier Series*. Proc.　London　Math. Soc. (2), vol. 18 (1919), pp. 141 - 162.

Zygmund, A.

1.　*Trigonometrical series*, (Dover Pub.)　Warsaw　(1935).